CULTIVATING
Agribusiness

To: Nancy McBride

With pleasant memories +
Best Wishes always!!

Mike Clinton
Sept. 2019

CULTIVATING
Agribusiness

DISCOVERING GLOBAL OPPORTUNITIES IN PACIFIC NORTHWEST SPECIALTY PRODUCTION

MIKE CHILTON AND BOB GRIFFIN

nunm
PRESS
PORTLAND, OREGON

NUNM Press 2016
Original impression published by NCNM Press (2015)
ISBN 978-0-9771435-9-7 (13 digit) • 0-9771435-9-7 (10 digit)

Managing Editor: Nichole Wright
Photographs provided by Mike Chilton

For permission to reproduce selections from this book, write to:
NUNM Press
National University of Natural Medicine
049 SW Porter Street
Portland, Oregon 97201, USA
www.nunm.edu

ISBN: 978-1-945785-10-8
Library of Congress Control Number: 2016949829

Production by Fourth Lloyd Productions, LLC
Cover and book design by Richard Stodart

Printed in the USA

This book is dedicated to all individuals and businesses
creating a future in the use of current and new specialty seeds
and crops supporting sustainable and economic livelihoods.

TABLE OF CONTENTS

PREFACE

This book intends to give credence to a personal vision for new and expanded agricultural productivity and technology for the Pacific Northwest. My vision foresees utilizing the existing natural and human resources of the Pacific Northwest as an expanded area of research, production, processing and marketing for specialty crops, especially those suited for cooler growing season environments. This enhanced capacity would serve the rest of the country and the world. This capacity could support a wide range of crops developed for proprietary end-use purposes, including: aromatic and special use oils, fibers, natural and pharmacological medicinals, plants used in arts and crafts sector, and the already significantly diverse and growing nursery, seed and food crops sectors.

There are a number of factors that encourage me to believe that this vision can be realized:

1. the inherent richness of agricultural production capacity in the Pacific Northwest;
2. the generally benign climatic conditions for both biennial and annual crops;
3. the availability of diverse choices of elevation and soil types to suit the needs of specific crops;
4. diverse climatic zones on the east and west sides of Cascade mountains;
5. relatively small owner-operated farms;
6. short distances to major hubs for both domestic and international transportation, and
7. educational institutions which have the capacity to become intellectual repositories for specialty crop production, processing and marketing.

I would hope this account of my agricultural endeavors will motivate potential entrepreneurs and existing agribusinesses to consider the opportunities that are available for the production of seed and specialty crops within the region.

I would also hope that my business experience will contribute case study material for courses in business and/or agriculture in tertiary and secondary education institutions in the Pacific Northwest.

Finally, I would like to reach farmers, extension agents and agribusiness firms with practical technical advice based on what I have learned in the course of years of field experience with farmers and various specialty crops.

The story we tell here did not follow in a straight line, but rather, it follows the circuitous route of my personal history evolving from deep and tangled roots. These include:

1. my youth as a farm boy on a small mid-western farm;
2. university training with a specialization in systematic and economic botany;
3. an extended stay in Southeast Asia working in agricultural development and related fields;
4. working for extended periods with people with widely varying cultural backgrounds, and reaching agreements in cross-cultural situations;
5. the reverse culture shock of returning to the United States after a long absence;
6. the long, meandering search for a new career;
7. and finally, the fortunate discovery of opportunities in Pacific Northwest agriculture.

Along the way, there were many anxious periods, but there were many periods of great pleasure as well that buoyed the spirit and created wonderful memories. My wish is to share some of my personal recollections in hopes that you will find them useful as you put together your future plans.

Mike Chilton
Turner, OR
September 2015

Co-Author Preface

I first met Mike Chilton in Pakse, Laos, in 1966. Mike was visiting a close friend from IVS days in Vietnam who had generously offered me a room in his house shortly after I arrived there. I was beginning a new job as regional manager for southern Laos of the USAID-funded Agricultural Development Organization. Mike had five years of IVS experience with Vietnamese agriculture under his belt and was beginning a new assignment in Thailand. I was just beginning my career in international development.

Over the next nearly five decades, we developed a strong friendship with shared interests in small farm agricultural development, Southeast Asia and entrepreneurship. (We also shared an affinity for the no-nonsense inventor, Bill Lear, creator of the Lear Jet and the 8-track audiotape.)

Mike and I began our working lives with direct exposure to the circumstances of subsistence farmers in Vietnam and Laos respectively. In the background of our experience was the Vietnam war.

Over the years we kept in touch by mail and often visited each other in places like Bangkok, Udorn and Songkhla in Thailand and Saigon in South Vietnam. (After Mike married Simone, a dinner invitation to their home was always favorably received.)

By the time the war ended, I had landed a consulting job on a UN regional project dealing with agricultural development. For Mike, the end of the war meant not only that Vietnam was no longer a place he could live but also that the thousands of jobs related to wartime development projects had simply evaporated. Furthermore, his dreams of agricultural production and marketing in Vietnam had vanished. After he returned to the family-owned acreage east of Springfield, Oregon, Mike ultimately looked at private sector jobs in the U.S. The main thrust of the book describes his adventures in Pacific Northwest agriculture.

When my eldest daughter moved to Oregon and opened a business with her husband, my wife and I had a new reason to visit the state and renew our friendship with the Chiltons in person. I had completed an MBA in the late 1970s. Though I never went into corporate employment, the course work engendered a lifelong interest in business news. Mike's business success stood out as it went so much against the grain. When Japan was no longer The Land of the Rising Sun, but in the words of an *Economist* cover story, "The Risen Sun", the U.S. was flooded with Japanese exports and seemed to be on a losing track in the battle for global economic supremacy. Japan was then in the 70s and 80s, as China seems to be today, the new colossus, destined to rule the world. Mike, nevertheless, had succeeded in producing a product that the Japanese were all too eager to import, daikon radish seed. His business' balance of payments was massively lop-sided for the USA. His story seemed made for packaging as a business school case study.

After Mike sold his business, we chatted off and on about documenting his business success. We came to realize that the business story was inextricably linked with his life story and really couldn't be told on its own. So what we have written here is really Mike's biography, a tale with many strands, many what ifs, many false starts and dead ends, but ultimately a successful coming together of experience and initiative that made for a significant one-man agribusiness success.

Bob Griffin
Honolulu, HI
September 2015

Acknowlegements

While the efforts of putting together any information always involve others, this writing has drawn upon many who have willingly given time and expertise to its construction. At the risk of omitting deserving others, here are a few who have been instrumental in this undertaking:

- the many cooperating growers over the years who have worked hard to produce a quality seed and helped to make Agriculture Services Corp., and later, Agricultural Alternatives, successful seed businesses;

- the many international customers who have cooperated with clarity and guidance, offering patience and understanding to assure that a successful seed crop resulted;

- fellow seed company owners and staff with whom we cooperated in the field and raised a vision for the creation and successful operation of the Willamette Valley Specialty Seed Association (WVSSA);

- Oregon State University (OSU) personnel including Bill Mansour, Bob Rackham and Bob McReynolds, who through the years provided objective guidance in the creation and operation of WVSSA and gave a sense of purpose to the growing importance of the Pacific Northwest specialty seed industry;

- Christopher Klemm, formerly Director of the Entrepreneurship Program at OSU, who has read and critiqued the original manuscript and provided useful guidance to its improvement;

- Dr. David Schleich, President of National College of Natural Medicine (NCNM), who has always provided encouragement and support for this book's publication, and to Nancy and Richard Stodart for their caring and tireless efforts in book design and layout; and

- my enduring wife, Simone, who has generously tolerated all the gyrations, changed plans, late hours, disappointments, and achievements that resulted in seeing this writing project to completion.

ONE

GETTING STARTED IN OREGON AGRICULTURE

In his book, Outliers, *Malcolm Gladwell debunks the idea* that geniuses are born as fully competent masterminds. In making his case for nurture over nature, he points to prominent innovators who made their biggest breakthroughs after long apprenticeships: Bill Gates spent weeks and months writing software programs in high school and college as part-time jobs and hobbies; Tiger Woods had a golf club in his hands from age three; and the Beatles' played 1,200 performances in Hamburg, Germany, prior to their emergence as Britain's greatest rock group.

On the basis of these examples, Gladwell formulates "the 10,000 hour rule" or the minimum investment of practical work experience required for a person to potentially produce extraordinary results in his or her chosen field.

While not everyone who does extensive on-the-job training will achieve remarkable results, it is practice, not genetic endowment, that is essential to "make perfect", at least according to Malcolm Gladwell.

My introduction to Oregon agriculture began in the mid-1970s. I had been working in Vietnam and Thailand for the previous fifteen years, initially as a volunteer with International Voluntary Services and later in both private and public sector positions. Most of these assignments had some kind of connection to agriculture and rural development. When the Vietnam War came to an end in April 1975, my employment opportunities evaporated. In July, my wife Simone, son Paul and I left Southeast Asia for San Francisco. The Government offered me a sixty-day temporary duty assignment in Portland, Oregon with the Army Corps of Engineers. The short-term job was supposed to be my transition to new employment. I did the sixty days with the Corps writing what was probably the first generation of environmental impact statements for their projects.

That assignment could have led to more work but we chose to move to a farm in Springfield, Oregon, that we had bought with my family a few years earlier. I was looking for a job, any job, and still at loose ends. It is difficult to express how one feels in such moments. I felt like a refugee in my own country. I needed some time to figure out what I wanted to do and what opportunities were available. One thing that I did know was that I had become disillusioned with the idea of working for the government. Eventually, the private sector would beckon.

GRAPES, BEER AND SNEAKERS

The emerging opportunities in the Oregon economy in the late 1970s and early 1980s were not obvious to me at the time. Chris Klemm, former director of the Oregon State University (OSU) program on entrepreneurship, reminded me of the ways that the Oregon economy was evolving in the years after I took up residence here.

While grapes had been grown for wine since the middle of the 19th century in Oregon, the modern era of viticulture began in the 1960s with new plantings by wine entrepreneurs who brought expertise and experience from the University of California, Davis, and from France. By 1981, an OSU survey found over ninety wine grape growers with a total of over 1200 acres in grapes, much of which was not yet bearing fruit. Two-thirds of that acreage was in grapes for Pinot Noir, Riesling and Chardonnay wines. Thirty years later, vineyards had expanded to over twenty thousand acres, of which over twelve thousand acres were in Pinot Noir grapes. Over fifteen varieties of wine grapes were being produced. The 2011 crop was 41,500 tons worth nearly $2,000 per ton.

As is the case with wine, Oregon has been producing beer since the mid-19th century. Henry Weinhard founded perhaps the most famous Oregon Brewery in 1852. Prohibition put the brakes on the production of all alcoholic beverages in the 1920s and 1930s. Though historical statistics on beer production are hard to come by, beer production slowly expanded over the decades until the state legislative approved the establishment of brew pubs in 1985. This act kicked off a large expansion of craft breweries. According to the Oregon Craft Beer website, Oregon now has 172 breweries that operate 213 brewing facilities with a total of 6,500 employees. Portland has fifty-five breweries and is known in some quarters as the micro-brewery capital of the world. Statewide, production of beer reached 1.4 million barrels of which Oregon residents consumed more than a third.

As beer production has expanded, so has the demand for hops. In the 1930s, Oregon was the largest hops-growing state in the country. There is still significant production in Oregon—over 5,500 acres in 2014—but the large majority of national hops production is in Washington state which currently has 29,000 acres planted. Growing hops is labor intensive, uses specialized technologies, and has relatively high start-up costs. These factors may explain why Oregon hops growers are mainly third or fourth generation farmers.

The third area of the Oregon economy that was taking off was not just sneakers, but the whole gamut of athletic shoes and sports apparel. Nike had started in the 1960s with Phil Knight selling Japanese-made shoes out of the trunk of his car at track meets. Along with University of Oregon track coach Bill Bowerman, they founded Blue Ribbon Sports (BRS) with the intention of building better running shoes. BRS became Nike in 1972 as the firm shifted from distribution of other manufacturers' shoes to the design and production of their own products. A Portland State University graphic arts student created the famous "swoosh".

By 1976, Nike had $14.1 million in annual sales. Five years later, in 1981, that number was $457.7 million. In the same period, earnings per share of the company's stock had risen from $.04 to $1.57. During the past three decades, the firm has created full lines of shoes and athletic apparel and expanded distribution around the globe. Sales in its fiscal year 2014 topped $27 billion. Roughly six thousand people are employed at its twenty-two-acre headquarters campus in Beaverton.

The knock-on effects of Nike's success are vital for the state's economy. Columbia Sportswear has it headquarters in Portland and Nike's main international competitor, Adidas, has its American headquarters there as well. Oregon has been discovered as a perfect place for outdoor gear firms. According to Will Blount, the President of Ruffwear, "Oregon is the ideal place to build, test and photograph outdoor products. Where else can you find snow, desert, mountains, forest, rivers, lakes and oceans all within a few hours of travel?"

In the pre-Internet age of the 1970s, I was aware of none of these promising avenues of economic development in the state. However, living on the farm in Springfield, in the middle of Oregon's most prolific agricultural region, re-kindled my interest in agriculture.

Learning Oregon Agribusiness

My desire to return to some kind of productive work was unrequited for many months. Finally, in January 1977, my 10,000-hour internship on the road to entrepreneurial agribusiness success really began. I was offered a job with a small, Salem-based firm called Agricultural Services (Ag Services). The company contracted local farmers to grow grass seed and sold them fertilizer blended for their soil conditions, along with other agricultural chemicals. My bosses at Ag Services, John Rutkai and Dave Amoth, each owned a quarter of the company. Van Der Have, a large Dutch seed breeder who purchased all of Ag Services' grass seed production, owned the other 50%. Despite its low profit margins, Dave and John focused their energies on the fertilizer blending operation. In the late 1970s, the firm's annual turnover was in the neighborhood of $6 to $7 million including both grass seed sales and the revenue from the fertilizer and chemicals segments of the business. The firm also had a very small vegetable seed business that wasn't attracting much attention from management. Much to everyone's surprise, this sideline took on great importance in the years to come.

When I began at Ag Services, John and Dave didn't quite know what to do with me. I swept the warehouse floors, trapped gophers and did some library research at Oregon State. After a few months with Ag Services, John asked me if I would like to take charge of the vegetable seed business. Over time, in what I look back on as my Gladwellian apprenticeship, I was to learn the ins and outs of contract seed farming, the value our international partner brought to the business, and the need for healthy margins in the often unpredictable seed business. I was on a path that would eventually lead me to the point where I would start up my own agribusiness.

Seed production is a staple of agriculture in the Pacific Northwest. East of the Cascade Mountains, high desert prevails. In the valleys of western Oregon and Washington, the climate is Mediterranean with mild, rainy winters and warm, dry summers. The soils are good, agriculture is diversified, and farmers are willing to try new crops and adapt their farming procedures as required. These conditions are highly favorable for vegetable seed production. The specific growing conditions in the fertile Willamette Valley are described in more detail in a pamphlet developed by the Willamette Valley Specialty Seed Association in cooperation with the Oregon State University Extension Service (see **Appendix One**). In addition to great growing conditions, the expansion of the Port of Portland and other transportation infrastructures in the area facilitate getting seeds to markets.

The vegetable seed industry, which became my specialty, started in the Pacific Northwest towards the end of the 19th century when a farmer named Alvinza Tillinghast successfully produced a crop of cabbage seed in Skagit County, Washington. Vegetable seed production expanded slowly over the following decades until the world wars, when the U.S. was cut off from European seed supplies. By the time of World War II, European seed production was in disarray. In response to the disruption of the industry, local and foreign seed companies set up operations in the Pacific Northwest. These businesses expanded rapidly to meet wartime demand, initially for cabbage, carrot, and sugar beet seed. By the 1970s, these seed operations had become permanent bases for American and European seed companies.

When I took over responsibility for Ag Services' vegetable seed contracting and sales, vegetable seed production was a small sideline for the firm and not a significant contributor to its bottom line. Despite the apparent insignificance of the seed operation, I was eager to take up this opportunity and actually had a good deal of relevant experience for the job. I knew something about seed production from my studies at Iowa State University. I had crop experience from earlier work experience in Vietnam. My grasp of plant taxonomy—the relationships among plants and their environment—was particularly important.

Start where you are.
Use what you have.
Do what you can.

Arthur Ashe

For example, knowledge of the pests and diseases of cabbage told me that radishes, which are taxonomically related, would likely suffer from the same production problems. Similarly, I had a firm grounding in plant physiology, the science of how plants grow. Thus, I knew that different crops reacted differently to temperature and hours of daylight. My knowledge of the market was also growing. I had learned that seed companies liked to diversify their sources of supply geographically in order to protect against a crop failure in any given region. Finally, I knew that Asia was a growing market for almost everything, and I was looking for a way to make use of my familiarity with the Far East.

THE DAIKON RADISH SEED BOOM

In 1978, I convinced my bosses at Ag Services to fund a business development trip to Asia to promote our capacity to grow vegetable seed. John and Dave had strong backgrounds in agricultural technology, but despite their partnership with Van Der Have, their international experience was limited. John and I spent three weeks visiting Japan, Korea, Taiwan, Hong Kong and Thailand. I used the services of Agricultural Attachés at U.S. embassies in these countries to identify local seed companies and make appointments for us. After three weeks, we came back to Oregon with two five-acre orders for two different varieties of daikon radish seed. The customer was Takayama Seed Company of Kyoto, Japan.

Daikon Radish

Daikon radish is a long white root crop familiar to patrons of Japanese restaurants as the shredded garnish served with sashimi and in soups. Takayama Seed had a growing domestic demand for radish seed and was willing to try us as a new seed supplier. The Japanese firm was unable to find tracts of land large enough for radish seed production in Japan as typical land holdings there were quite small. Mr. Ono of Takayama gave us a contract, probably knowing full well that we had never grown daikon seed before.

Blissful in our ignorance of the vagaries of daikon seed production and the related seed quality requirements, we plunged ahead. The business model was simple and already commonly understood by grass seed growers in Oregon. In order to be able to contract for a grower's seed production, our first priority was to identify and secure a market for that production. In the case of daikon seed, we had found a small market. Our Japanese buyer agreed to buy as much seed as we could produce on pre-agreed acreage for X$ per pound. Next we had to find growers who would agree to grow a certain number of acres of daikon seed that we would buy for Y$ per pound, a price that would give us a comfortable profit.

The agreement included standards on germination rates and purity for the grower's seed. X minus Y would be our profit margin, which we locked in by means of a grower's contract on the production side and a buyer's contract on the marketing side. To make this business model work, we had to find growers who were willing to produce for Y dollars per pound and who could meet our buyers' specs for germination and purity.

The business model sounded simple enough, but we immediately encountered difficulties. The first seed production problem was a lack of interest from farmers. Eventually, we found Dave Fuller, a farmer who thought we offered a good deal and was willing to take a risk with a new crop. We contracted Dave for five acres of seed. He became our first and only contractor in our initial growing season. Secondly, we found that growing was easy: daikon grew very quickly. But harvesting proved to be difficult, really difficult. The trick was to get the daikon seed out of its pods without damaging it. Traditional harvesters destroyed much of the seed or failed to break it out of the pods. To deal with unopened pods, we tried running them back through a burr mill by hand with limited success. Nevertheless, despite the harvesting problems and some contamination from wild radish that grew naturally in Dave's fields, demand in Japan was such that Takayama took all the seed that we could deliver. However, due to the additional threshing expenses, we only broke even in our first year.

The second growing season came around quickly. The news had spread by word of mouth in Japan that there was a new source for daikon seed in Oregon and we had some Japanese visitors. Though we added only one new Japanese customer in 1979, we had seventy-five acres of daikon seed in our order book—fifteen times what we had in the ground in 1978.

Again in the 1979 season, harvesting was our biggest concern. We tinkered with harvesting equipment, adding rubber padding and rollers to cushion the threshing process. Ultimately, we contracted a friend with an engineering background, Hank Warkentin, to fabricate the first set of rollers and pads. The padding and rollers greatly improved our seed quality. These modifications were particularly helpful if the crop had a high moisture content. Over the next few years we had Hank make twenty more sets of rollers. We rented them to our contract farmers and required their use for harvesting as a condition of our grower contracts.

While we were struggling with traditional combines, International Harvester began to introduce a rotary combine in our area. The new rotary combines threshed daikon seed more gently and efficiently than older harvesters, eventually becoming the main choice of local seed growers for their harvesting needs. When the new harvesters were introduced, their owners didn't have enough work for the machines and were happy to do contract harvesting on our daikon fields. In any event, the rotary combines didn't need rubber rollers. Our tinkering with the old harvesters had solved our harvesting problems for a couple of seasons, but the introduction of rotary combines provided the long-term solution.

The trajectory of our daikon seed production over the next few years was almost vertical as can be seen from the following table:

AGRICULTURAL SERVICES DAIKON SEED ACREAGE 1978-1983

Year	Acres Planted
1978	5
1979	75
1980	170
1981	500
1982	1400
1983	2500

Yields averaged around one thousand pounds per acre, but the top growers were able to produce 1600-1700 pounds.

I tried to make arrangements with our growers as simple as possible. Over time, we reduced the contract to a one-page document that focused on the mandatory production techniques, the seed quality and the moisture standards. We made an effort to take the legalese out of our grower contracts. Samples of Marketing and Production Agreements, which I subsequently further refined and used in my own business, are presented in **Appendix Two.**

What we hadn't understood when we got into the daikon seed business was that the natural food craze was sweeping Japan in the early 1980s. Daikon sprouts had become a prominent feature of "healthy" meals. This phenomenon was the Japanese equivalent of getting alfalfa sprouts on a sandwich at a health food deli in the U.S. Our business development trip to Japan had gotten us into the daikon seed business and we developed a strong capacity to deliver seed to our buyers. Catching the wave of the daikon seed boom was a great piece of good luck.

The consumption of daikon radish was holding steady in Japan, but the consumption of daikon sprouts was exploding. We sold our seed output to Japanese seed companies who in turn sold it to sprout growers. Two-thirds of our seed output was passed through to sprout producers. Over time, the sprout producers demanded higher quality seed and set specifications for uniform size, "vigor"—the germination rate—and moisture. We rode the daikon sprout wave realizing healthy margins on sales to our Japanese customers.

Over time, Japanese seed companies had tried to diversify their sources of daikon seed supply. They started production in Italy at comparable latitudes with similar weather conditions to Oregon but growing on the small fields that were available there resulted in wide variations in seed quality— particularly in germination rates and the vigor of sprouted seed. In Oregon we had the advantage of large fields that yielded a uniform product.

The year 1983 had been wetter than average and the moisture had affected our production. There were predictable problems with harvesting including more damage to the crop and a higher percentage of non-threshed seedpods. Germination rates were also lower. When confronted with this problem in the grass seed business, seed companies simply blended moist seed with dry seed. While this practice worked for grass seed, it didn't work for daikon seed. Blending moist and dry seed noticeably reduced seed quality. Japanese sprout growers complained to the Japanese seed companies who had bought our seed and the seed companies immediately complained to us.

I wasn't a party to the decision to blend wet and dry daikon seed. I was unhappy with this unethical short cut as I knew how important it was to maintain the confidence of our buyers. The breach of confidence had badly damaged Ag Services' reputation with Japanese buyers and I began to wonder how much longer I could continue with the firm.

T W O

RUNNING MY OWN BUSINESS

AGRICULTURAL ALTERNATIVES, LTD.

*I*n *1985, I established Agricultural Alternatives, Ltd.,* my new company, to work with farmers to produce seeds and crops on a contract basis, much the same as I had done with the vegetable seed operation at Ag Services.

EVENING PRIMROSE, AN OIL CROP

Evening Primrose (*Oenothera* spp.)

The American Indians had discovered that primrose root could be eaten for nourishment and that the application of its oil aided the healing of wounds. Today, there is a large volume of research supporting the use of the oil for a variety of medicinal purposes.

Ag Services had been successfully growing evening primrose since 1977. Primrose is primarily a commodity crop pressed for its oil and secondarily a seed crop. Ag Services began production of evening primrose with twenty-five acres in the Skagit Valley. In 1978, production shifted to the Willamette Valley, again with twenty-five acres planted. From 1979-84, production was relocated again to east of the Cascades in Washington and central and eastern Oregon, areas that were climatically more favorable for growing primrose. Efamol, a British company, was our sole buyer.

In 1985, after I had set up Agricultural Alternatives, I visited Efamol's facilities in England and found that they were eager to find additional sources of primrose production. On my departure from Ag Services I had agreed that I would not compete with their primrose production in Oregon. As Efamol was interested in diversifying its sources of supply, I established my own new primrose production under Agricultural Alternatives in Virginia and North and South Carolina. Efamol went so far as to subsidize me to the tune of $20,000 per year for four years to get my production capacity up and running.

The primrose business in the eastern states never really got off the ground. The farmers there were used to growing corn, beans and tobacco. We felt tobacco growing was closest to primrose cultivation because we were hoping to apply tobacco's transplanting practices. But primrose margins were not nearly as good as vegetable seed margins, so in 1989 I returned to the Pacific Northwest as a base for vegetable seed and primrose production and closed down operations on the east coast.

While Efamol focused on the breeding development of readily producible varieties, i.e., an increasingly determinant seed set, reduced shattering, and higher gamma linolenic acid content, we focused on agronomic problems to enable more economic production. This work included techniques for better and faster seedling emergence, more effective and less costly weed and pest control, and optimum harvest protocols. We were rapidly converting the plant from a wild North American native plant to a commercially productive crop.

Agricultural Alternatives had successful primrose crops until 1997, after which time production migrated to other regions of the world with lower production costs, mainly China. Our best years for primrose were the early 1990s in the Pacific Northwest. In the peak years our contract growers produced 800,000 to 1,000,000 pounds of oil seed.

Vegetable, Flower And Herbal Seeds

The Willamette Valley Specialty Seed Association describes the valley as "the perfect environment for seed production" with "fertile soils and a year round temperate climate". Rainfall, which averages 1000 mm annually, is skewed towards the winter months. July and August receive little precipitation creating generally excellent harvesting conditions. Seed production in the Pacific Northwest dates back to the turn of the 20th century, providing a long history for current growers.

My first five years as an independent agribusinessman were, frankly speaking, slow. In addition to the primrose business—the subsidy from Efamol was keeping us afloat—we began multiplying specialty vegetable seeds for seed companies. We also received a large order for peppermint leaf for tea to replace lost production due to the Chernobyl nuclear disaster in 1986 that contaminated growing areas in Europe. Peppermint was already being grown in the Willamette Valley for its oil. We contracted peppermint growers to produce dried leaf instead of oil which we handled much like hay, drying the plants in windrows, baling the crop and running it through a stationary combine to separate leaves from stems. We exported about fifty tons of tea leaves that year. The following year the buyer found other sources for its peppermint tea, so we moved on to other opportunities. Peppermint provided a lesson on the need for nimbleness in the seed business. The demands of the marketplace were often fleeting.

Peppermint (*Mentha piperita*)

By 1988, three years in operation, Agricultural Alternatives' sales had reached $200,000, all of which were from exported product. Continuing our export market ventures, we eventually grew nearly thirty-three different vegetable seeds. The major vegetable seed crops, that is, those with large acreages or high yields included daikon, red radish for greenhouse growers of radishes, and seeds for cabbages, carrots, Brussels spouts, Chinese cabbage (choi sum), Chinese kale (kailan), Chinese mustard (bok choi), collards, other Chinese mustards, bunching onions, peas, rutabagas, turnips, kale, parsnips, spinach, sugar beets, and Swiss chard. These seeds were exported to European and Asian seeedhouses.

We also grew seeds for twenty-eight different kinds of flowers. Our main flower varieties were: calendula, California poppy, candytuft, clarkia, annual chrysanthemum—used as in ingredient in sukiyaki—cosmos, dame's rocket, primrose, meadowfoam, and wallflower. Like primrose, meadowfoam was grown as a seed crop for oil extraction.

Finally, we grew roughly thirteen herbals, the most important of which were: burdock for the Japanese market, borage, cilantro (or coriander), chives, dill, parsley, peppermint, and echinacea.

Burdock (*Arctium* sp.)

In regard to these minor seed crops, my familiarity with botanical relationships was particularly helpful because we had to develop our own agronomic protocols. My botany training had taught me that protocols for more common plants could be applicable to plants within the same families. The application of this knowledge was critically important to the growth of these lesser-known crops.

A number of these specialty seeds were only grown for two or three years. We often found that the requirements of our buyers varied from year to year as market demand shifted. On the other hand, we may have had mediocre agronomic results with a crop, such as Chinese parsley, and decided on our own to stop growing it.

WORKING WITH GROWERS

What a great pleasure it was to meet and work with growers in the Pacific Northwest in general and the Willamette Valley in particular! These growers, at least the ones who agreed to work with me, are extraordinary men with extraordinary families, and I am proud to continue to call them my friends to this day.

Bob McReynolds, a longtime horticultural specialist with the Oregon State University Extension service, has described the types of farmers he has encountered in the course of his work:

First, there are newcomers who see agriculture as a way to get rich quick. They jump on the bandwagon of the latest crops and technologies regardless of the production and marketing realities.

Second, there are those whose working life is consumed by constant confrontation with seed companies, mainly over prices and contract terms. A conversation with one of these growers leads inevitably to a recitation of the latest conflict with an "evil" seed company or its field agent who was ruining farming for everyone. (There's often little love lost among growers and seed companies. Every grower can tell you a story about how a seed company had cheated him or his neighbor.)

A third group of farmers might be called "old school". They grow low risk crops and are content to "ride tractor" as they go through the motions of traditional farming.

Finally, there are the true entrepreneurs—the smartest, most practical and most likely to survive in the complex, changeable environment of free market agriculture in the Pacific Northwest. I sought to work with members of this last group.

Entrepreneurial farmers have a skill set that enables them not only to survive, but also to succeed. They are able to recognize opportunities in the marketplace, are able to calculate the risks and rewards, and are willing to take risks. They understand their operating costs and are able to control, if not reduce, them. They adopt innovative business practices not only to reduce costs, but also to improve efficiency. Finally, they appreciate that the introduction of any changes in crops or technologies should be incremental and carefully monitored.

I have a strong preference for growers with an entrepreneurial bent. In the Willamette Valley in particular, the extent to which farmers are engaged in agriculture as a free enterprise activity is remarkable. In other words, the growers there do not seem to be benefiting from the federal Farm Bill or other state or federal agricultural subsidies. In conversations with Willamette growers over the years, they have explained how they regularly scan the markets for cropping options, consider candidates for the diversification of their production, take into account weather, growing times, crop rotation benefits, and labor needs, and consider competing offers from seed companies, food processors and fresh markets. These growers are computing a large number of variables on a rolling basis in order to make their agribusiness decisions. What is interesting is that they don't seem to be taking into account government regulatory activities or agricultural subsidy programs—unlike their brethren in the Midwest.

Are these individuals the last free market capitalists in agriculture?

I asked Bob McReynolds of Oregon State University for his opinion on this question, based on his years of experience in Valley agriculture.

> I would say you're correct about your understanding of the growers here. I doubt that they really want to have the government involved in their decisions. Many of the subsidies are offered because the growers can't compete in the world without help and the government needed to support the family farm for political reasons. The crops grown in the valley are considered specialty crops or minor crops in the greater scheme. Niche crops might be another term. The growers have found these niches that they can compete in and do quite well. Other areas of the U.S. lack some of the competitive capabilities or advantages that Oregon offers, such as the moderate temperatures in the Pacific Northwest and the semi-Mediterranean climate with limited rainfall in the summer, just to mention a couple.

Gary Cooper, who runs his own seed contracting business from his home near Jefferson, Oregon, agrees. In his view, "the productivity of the land in the Valley with its high yields encourages independent production. It creates opportunities for small business." He notes further that big seed companies are expanding in the Valley and that this is squeezing the smaller seed contractors.

There's also the fact that Valley seed producers plant large acreages of grass seed, hardly a sympathetic candidate for subsidies, especially in comparison to crops grown for food or energy.

While the natural endowment of the Valley may establish the baseline for economic opportunities for growers, I see the genius of these entrepreneurial farmers and their old-fashioned work ethic as the key factors in their business success.

Selling Agricultural Alternatives, Ltd.

By the mid-90s, Agricultural Alternatives averaged around a million pounds of seed production annually. Annual sales in the period 1989-98 shown on the table below were hardly consistent but we were able to remain profitable in the leaner years by keeping our overhead under control.

In the crop year 2000, we had seventeen varieties of seed crops in production worth $1.6 million based on the contracts in hand. The most important crops at that time were radish, cabbage and various onion varieties. Gross margins varied with the seed variety but the average for our main crops was around 25%.

In 2001, we had signed contracts that had the potential of yielding $2.5 million in gross sales. This would have been the highest gross revenue in the life of the company. This potential made Agricultural Alternatives very attractive and I took advantage of the opportunity to sell out in February of that year.

Agricultural Alternatives Gross Sales 1988-2000

Year	Gross Sales in US$ millions
2000	1.61
1999	1.41
1998	2.15
1997	1.44
1996	0.94
1995	0.85
1994	1.78
1993	2.16
1992	1.80
1991	1.10
1990	0.53
1989	0.30
1988	0.20

Mike Chilton's home base of operation for Agricultural Alternatives, Ltd., from its birth in 1985 to its close-out in 2001. Photo by Norma Griffin, 2015.

THREE

FACTORS AFFECTING OREGON AGRIBUSINESS

RESPONSIVENESS TO CHANGED CIRCUMSTANCES AND OPPORTUNITIES

A number of inter-related factors affecting the present and future of Oregon agribusiness are briefly discussed below. Agricultural trends in the Pacific Northwest in general and the Willamette Valley in particular are always in flux. For decades, growers produced crops for canneries, but around 1984, the canning industry in the Pacific Northwest collapsed. Times had changed. Consumers were switching from canned to fresh or frozen fruits and vegetables. Competition from suppliers in other areas of the U.S. and from foreign countries was also eating away at the market for canned products. The impact on local production was felt almost immediately.

For example, Brian Parker, who runs a 370-acre family farm near Junction City, Oregon, had his entire farm in sweet corn for canning in 1984. As cannery demand for sweet corn declined, he was forced to search for alternative crops. Like any smart entrepreneur, he focused his search on crops that had higher margins. He tried cauliflower and cabbage for the fresh market. Those crops proved to be more work than they were worth. Later he took contracts from seed companies to produce foundation grass seed. By the early 1990s he had diversified into a small acreage of specialty seeds and added a seed cleaning capacity for his own crop and for contract cleaning of other growers' seed production. Following this successful diversification, Parker says that now he must turn down seed contract offers as the offers exceed the production capacity of his family farm. As Parker's experience suggests, there already may be available opportunities for growers who are willing to try specialty seed and other new crops.

THE CYCLICAL NATURE OF AGRICULTURAL PRODUCTION

Agricultural production has its own variant of the business cycle that is less a trend than a fact of life in farming. "With cabbage, I get one good year, two average years and one total bust," says Wilbur Klopfenstein, who with his extended family, farms grass seed, flowers, specialty seeds and blueberries near Salem. Seed contractor Gary Cooper explains simply that the seed business is very competitive.

Change before you have to.

Jack Welsh

"Good prices mean higher production. More production means lower prices, tighter margins and ultimately lower production."

Farmers can't escape the basic laws of economics in a free market environment and are thus always on the lookout for the next opportunity.

SPECIALTY SEED DEVELOPMENT

Grass seed is the elephant in the room in the Willamette Valley with farm gate sales estimated at over $300 million in 2012. Specialty seed production is rising fast with sales at over $50 million for the same year. Grass seed is grown extensively over large acreages. Seed companies give contracts to growers and leave them alone to produce the crop. Specialty seeds are grown intensively on small fields, usually 5 to 10 acre plots, and require more management and labor. Specialty seed companies also provide a high level of monitoring and supervision throughout the growing season, including farmer training, as a way to prevent production and processing problems. Specialty seed crops are higher value crops. However, higher reward usually entails higher risk. A grower who chooses to try a specialty seed crop needs to be able to tolerate the associated risk. Nevertheless, despite the risks, the rewards from specialty seeds and other specialty crops are sufficiently appealing to local growers that they are willing to diversify a part of their production to them.

MAINTAINING VARIETAL PURITY

Preserving varietal purity was a critical production goal and one of the key factors in maintaining a positive reputation with our buyers as seed acreages increased. I seldom had to think about issues that involved other seed companies or growers, but occasionally it became necessary for agronomic reasons. When I was still with Ag Services in 1979, we were expanding our production base with several 5 to 10 acre plantings of cabbage and other cabbage-related vegetables, all members of the mustard or Cruciferae family. Several of these vegetable seeds could genetically cross. Even if they didn't cross, their seeds were essentially indistinguishable from each other.

> *Coming together is a beginning.*
> *Keeping together is progress.*
> *Working together is success.*
>
> Henry Ford

In either case, there is a serious problem with physically mixing seeds within the produced crop, either from unwanted volunteer plants in the field or residual seed carryover from harvest and handling equipment. My botany training came to the fore very early to enable me to recognize the severity of these problems. From a marketing perspective, we had learned very quickly the importance of purity and varietal integrity of any production undertaken for our international buyers. The expectations were very high in terms of the seed quality and purity. Contracting with these customers would be unsustainable if seed quality deteriorated.

Our cabbage crops were in danger of cross-pollinating with other cabbage varieties in nearby fields being grown by farmers contracted by other seed companies and vis-versa. As we began to increase our radish acreages, similar problems of cross-pollination were to occur not only with planted varieties but also with endemic wild radish plants. While weedy species of the mustard family did not often cause inter-crossing with vegetable species, the physical contamination of weedy seeds in the finished product was extremely undesirable and always a critical issue. Good field management almost always resolved this problem.

A few seed company representatives, myself included, started to discuss ways and means of protecting our crops from cross-pollination. Bill Mansour and later Bob Rackham of the Oregon State University (OSU) Extension Service joined us in these discussions. These two men provided substantive expertise and a neutral perspective that greatly facilitated our discussions. Eventually, isolation guidelines that set minimum distances for separation of seed crops were drafted and agreed to by the participating seed companies. An arbitration mechanism was included in the isolation protocol, but over the years it was invoked only once. What started out as an informal working group became, in 1980, the Willamette Valley Specialty Seed Association (WVSSA), an organization whose goal is to promote quality seed production. The organization has developed isolation guidelines, physical separation distances and mapping procedures for all of the common seed, vegetable and flower crops. The WVSSA is an admirable example of practical problem solving in agriculture by the development of cooperative relationships among growers and seed companies. OSU facilitated the arrangements. Neither new state laws and regulations nor interventions by the Oregon Department of Agriculture were required.

THE DISPUTE ABOUT CANOLA

The bit of history above is instructive in view of the controversy—"Canola chaos" in the words of Bob McReynolds—that has been percolating over the past few years. Canola refers to both a plant and the oil produced from its seeds. In the 1970s Canadian scientists bred canola naturally from the rape plant to enhance its nutritional characteristics. Today, the crop is grown extensively in the U.S. and Canada. Canola oil has many common uses: as an edible oil, as a component in lubricants and as an ingredient in bio-fuels. Subsequent research created a genetically modified version of the plant that is resistant to herbicides. The vast majority of canola production in the U.S. and Canada is now the pesticide-resistant variety.

The controversy came to a boil in August 2013, when Oregon's Governor John Kitzhaber signed into law HB2427 relating to the growing of canola in the Willamette Valley. The law effectively redefines the Willamette Valley

Protected District whose regulations had strictly forbidden the production of canola in the Valley. The Department of Agriculture website explains that the new law does the following:

- establishes a moratorium on the growing and raising of canola within a defined protected district of the Willamette Valley.
- authorizes growing not more than 500 acres of canola within the protected district for the purpose of carrying out the Oregon State University research that was funded and authorized by the 2013 Oregon State Legislature.
- allocates $679,000 to Oregon State University to carry out the research proposal.

OSU is supposed to report the outcomes of the research study and present its findings no later than November 1, 2017 to the Oregon State Legislature.

On the surface, the dispute about canola appears to be an argument between winners and losers within the agricultural sector in the Valley. On one side are grass seed growers who recognize the value of canola as a soil enriching rotational crop that, unlike other rotational crops, can be sold at a profit. On the other side are the specialty seed growers who understand the potential damage that canola, another brassica, can cause to seed crops. Dig a little deeper and you will find canola supporters include environmentalists promoting renewable energy and politicians who want to be seen as supporting "green" initiatives. The larger universe of canola foes incorporates groups and people who are opposed to genetically modified crops in general.

The Willamette Valley Specialty Seed Association (WVSSA) has explained its opposition to canola production in the Valley on its website. (See **Appendix Four** for the complete text.) The crux of the Association's argument is reprinted here:

The WVSSA contends that canola is damaging as a crop, weedy volunteer, and host to insect and disease pests negatively impacting specialty, clover, and grass seed producers and related interests, such as fresh market vegetables. Evidence is available. For example, Crucifer (*Brassica* species and radish) seed production areas in Europe have been ruined and lost indefinitely, particularly in Denmark, the UK, and France. This is long-term damage that is not easily reversed or negated. Depending on the specific crop, canola variously affects seed or food quality. Examples include, but are not limited to, unwanted cross pollination, undesirable weed seed content, insect pests such as pollen beetles and cabbage root maggots, and plant diseases such as those caused by the Sclerotinia fungus.

In other words, the seed producers are confident in stating that canola production is a menace to seed crops and fresh market vegetables; that the damage is long-lasting and difficult to reverse, and that there is no need for further research, given the evidence that is available from Europe. Whether genetically modified or "natural" canola is planted is immaterial. The seedsmen see the new law as a triumph of politics over science. They are deeply disappointed that the ODA has not made the scientific case on the basis of existing evidence more persuasively. There is a general fear that the 500 acres of canola now allowed in the Valley will be the tipping point for large-scale contamination and crop damage of seed crops.

If there is a lesson here, it is perhaps that politics and agricultural science are uncomfortable bedfellows. We were able to head off the problems of cross-pollination among cabbage and radish growers in the Valley by cooperative action based on common interests. The canola situation involves competing interests with the state attempting to craft a political compromise. The results so far are not encouraging. The time has passed when growers and seed companies with technical support from the Extension Service could come up with mutually agreeable solutions to agricultural issues. Politics and the prospect of financial gain are in the driver's seat now.

INTERNATIONAL MARKET RISK

The Willamette Valley has a worldwide reputation as an excellent place for seed production. According to the OSU Small Farms Program:

> The Willamette Valley is one of only five areas in the world where vegetable seed can be successfully produced. The unique growing

characteristics, such as weather and climate, contribute to the success of its specialty seed production. Because of these important characteristics seed production must be protected so growers can continue to provide high quality seeds to growers in the United States and the rest of the world.

Above, we have noted the potential dangers of failure to geographically isolate canola production from specialty seeds and crops. International seed buyers are sticklers on contractual standards for seed purity, that is, cleanliness and lack of contamination from other seeds. These standards are minimally 99% purity. Proximate canola production introduces the likelihood of volunteer canola plants in seed crop fields. Canola seeds will contaminate seed crops and furthermore they are virtually impossible to remove in the seed cleaning process.

Several European countries as well as Japan and Mexico, inter alia, and local governments in those countries have taken steps to ban genetically-engineered foods or seeds for planting. These actions may have been in response to Internet-generated hysteria or they may merely be protectionist measures to limit the impact foreign competition on local farmers.

Overall U.S. economic policy is another international market risk variable. Following the financial crisis in 2008, U.S. fiscal policies to promote economic recovery have relied on deficit spending funded by international borrowing. Monetary policies have artificially kept interest rates near zero. The results of these policies have increased U.S. sovereign debt dramatically as well as the size of the Federal Reserve's balance sheet. For several years, one consequence of these policies was the maintenance of a weak dollar *vis a vis* many of our trading partners. The weak dollar made our seed exports attractive to international buyers and this was good for the seed business. More recently, our major trading partners have pursued policies designed to weaken their currencies. Consequently, the U.S. dollar has strengthened and thereby made seed exports less competitive.

WHITHER THE FAMILY FARM IN OREGON?

While the growing conditions in the Willamette Valley may present a large inducement to new entrepreneurial entrants in agricultural production,

Bob Rackham and Bill Mansour, retired OSU extension men, point out a worrisome structural trend that is affecting the future of the family farm. The average farmer is fifty-six years of age and his adult children are tending to move off the farm, frequently seeking a university education and jobs in the city. There has been a national trend of small farm consolidation over the past three decades. In Washington and Oregon, average farm size has increased over 50% during that period. Though farming today is not the endless drudgery that it was before mechanization, it is still hard work with long hours outdoors regardless of the weather. The present agricultural labor force is comprised largely of migrant Mexicans who have become an essential element of farming in the Pacific Northwest today. Thus, while family farms remain critical components of agricultural production, smaller farms are getting swallowed up and family farms appear to be becoming more corporate.

Genetically Modified Organisms (GMOs)

Chris Klemm, a former OSU Business School professor, recently asked me, "What's all the fuss about GM (genetically-modified) foods?" This perplexing question appears to be another unfortunate intersection of politics and agriculture.

GM crops are not grown in the Willamette Valley since GM versions of the crops grown there have not yet been developed. Canola is mainly a GM crop, but the issues discussed previously regarding canola are not GM issues per se, as cross pollination and pest and disease would occur regardless of whether the seed was GM or natural. Also, though canola oil can be used as cooking oil, its production in Oregon is being touted for use in bio-fuels, so it wouldn't be used as a food.

In my time in the seed business, GM crops were not an issue. However, as a botanist, I have followed with interest the developments in agricultural biotechnology that began several decades ago and have resulted in large gains in agricultural productivity.

The continuing introduction of genetically modified crops is understood in some circles as being the latest wave of the green revolution—green referring to agricultural productivity not, as at present, meaning "environmentally sustainable". Norman Borlaug sparked this revolution as the lead researcher on cross-breeding of wheat in Mexico in the 1950s and 60s. Borlaug produced improved varieties of wheat with dramatically increased yields and incorporated disease resistance. The wheat crop in Mexico exploded and the improved varieties were exported to India and Pakistan which experienced similarly large increases in output, much to the astonishment of the believers in the theory that population growth would cause massive starvation in the near future.

Borlaug did his plant genetics in the open fields of experiment stations. In the early years of his research, his team did thousands of crossings of wheat varieties. Today, research on plant breeding has largely shifted from the experiment station plot to the laboratory. Seeds are modified in the lab using the techniques of genetic engineering. There is broad agreement in the scientific community that food produced from GM seeds is safe. Data from consumer groups and nutritionists that oppose GM foods has not been as scientifically persuasive. Borlaug himself became a vocal advocate for genetic modification of crops particularly as a means of meeting the demand for food in poor countries. According to Wikipedia (key word: Norman Borlaug), Borlaug believed that the genetic manipulation of organisms was the only way to increase food production as the world runs out of unused arable land.

Furthermore, in a review of Borlaug's 2000 publication, *Ending World Hunger: The Promise of Biotechnology and the Threat of Anti-Science Zealotry*, Rozwasdowski and Kagale argued that Borlaug's warnings were relevant in 2010:

> GM crops are as natural and safe as today's bread wheat. ....[Borlaug] also reminded agricultural scientists of their moral obligation to stand up to the anti-science crowd and warn policy makers that global food insecurity will not disappear without this new technology, and, ignoring this reality, global food insecurity would make future solutions all the more difficult to achieve.

This consideration brings me back to Chris's question, " What's all the fuss?" First, we have to recognize that genetic engineering and genetically-modified crops are disruptive technologies. Such technologies typically create new groups of winners and losers. Furthermore, the pattern of adoption of new agricultural technologies in general follows a bell curve beginning with innovators and early adopters and ending with late adopters and laggards.* Thus, at any given time, in the groups affected by the technology there may be adopters and non-adopters.

Farmers who suffer losses and perceive that they may suffer losses include, for example, those whose non-GM crops may be cross-pollinated by a GM crop. Another group of growers who may suffer are those who misuse or violate the terms of their contract with a GM seed company. Monsanto, for example, has sued farmers who have held back seed from their GM crop for subsequent planting or third party grain processors who have re-sold GM seed purchased from farmer A to farmer B for planting.

Then there is the group Borlaug identified as the "anti-science crowd". Opponents of GM foods are often not familiar with agriculture, and their claims can lack scientific rigor. Their positions include, for example, assertions about the dangers of eating GM foods ("frankenfoods"), about corporate agribusinesses concealing information about their activities, about excessive use of herbicides on GM crops, and about damage to the environment, inter alia.

Relevant to this recurring and robust debate are comments from the Food and Drug Administration (FDA) Fact Sheet *Food from Genetically Engineered Plants*:

> Foods from genetically engineered plants must meet the same requirements, including safety requirements, as foods from traditionally bred plants. ...Foods from genetically engineered plants intended to be grown in the United States that have been evaluated by FDA through the consultation process have not gone on the market until the FDA's questions about the safety of such products have been resolved.

In a recent interview in the Wall Street Journal, Monsanto Chief Operating Officer, Brett Begemann, added:

> The biotech-derived products that we eat are the most highly tested and regulated components in what we consume. A new

---

* Everett Rogers, *The Diffusion of Innovations*.

seed must be reviewed by the Department of Agriculture. Then there's a voluntary check from the Food and Drug Administration. If the GM seed includes insecticides or pesticides, as most do, the Environmental Protection Agency gets a look. It takes about $100 million to get one seed from discovery to market. Crops that are bred conventionally, on the other hand, undergo no government testing. None.

Mr. Begemann points out that supporters of GM crops include the FDA, the World Health Organization and the British Royal Society, all of which have declared GM crops as safe as conventional crops.* In addition to the issue of food safety, the *Wall Street Journal* interviewer noted the financial and economic benefits for farmers who plant GM seeds: In 2011 farmers earned $19.8 billion added economic benefit from GM crops, according to a 2013 report by the U.K.-based PG Economics. Genetically modified seeds are more resilient, yield more crops on less land and require less labor.**

Anti-GM activists in Oregon succeeded in getting an initiative on the ballot for November 2014 on the labeling of foods with GM content. The initiative was defeated. However, had it passed, there are GM proponents who predict that expensive and unnecessary regulation would have been put in place that may well have precipitated further protests and proposals for restriction of GM agriculture. As well, the simple requirement for labeling would have required analysis of every food product for GM content. It is unclear who will have had to bear the cost for such potential compliance: Producers, distributors, or retailers? Specifically, the argument continues, who would bear the cost of analysis? Which agency or group would determine standards as to what constitutes GM content? At the same time, such potential legislation would have to include an enforcement procedure and related administrative arrangements to validate that labels were in fact true in accordance with definitions and regulations prescribed by the regulatory body.

There are many who contend that the bottom line for specialty seeds is that so far, with the exception of canola, the GM issue is on the periphery of the business. GM seeds at present are available only for field crops that

---

* Kate Bachelder, "Meet Mr. Frankenfood," *Wall Street Journal*, August 23-24, 2014, p. A11.
** Ibid, p. A11.

are farmed extensively. Experienced agriculturalists wonder if GM seed will ultimately be available for specialty crops. The high cost of development will likely discourage research on GM seeds for specialty crops for the foreseeable future.

FOUR

BUSINESS LESSONS LEARNED

I have been out of the seed business for fourteen years now, but the twenty-four years I spent with Agricultural Services and my own company, Agricultural Alternatives, are never far from my thoughts. Many of my clients became my friends and we are still in contact.

Of course, the agribusiness climate has evolved a great deal since my first daikon radish field was planted in 1978. The Skagit Valley in Western Washington was then the focus of specialty seed production in the Pacific Northwest. There were only two or three companies doing specialty seed production in the Willamette Valley in Oregon. Since then, seed production has greatly expanded in Oregon. This growth is reflected in the fact that there are now over thirty local and international seed companies in the Willamette Valley Specialty Seed Association that was set up in 1981. The Pacific Northwest in general is now recognized as an important global center for cool season specialty seed production. Over the past four decades, the business environment has evolved so that today an aspiring agribusiness developer faces a market that is more competitive and more complicated, both locally and internationally.

I often ponder why the business worked out well for me and speculate whether a newcomer, starting from today, could do what I did. So to elaborate on this thought, I'd like to present some of the things I've learned —after a few years of reflection. I believe these lessons, presented below, remain valid today.

THE IMPORTANCE OF BOTANICAL TRAINING

My university education in botany provided an excellent foundation for all my work in agriculture. Knowing how plants grow—their physiology— and how they are related—the relevant taxonomy—have been eminently useful. For example, I could infer that the techniques for growing mustard would likely work for cabbage as they are both taxonomically classified as *brassicas*. Similarly, I recognized the dangers of cross-fertilization from canola as well as the dangers related to the spread of plant diseases.

I recall that botanists at Iowa State liked to say that "good agriculturists should also be good botanists". I never fully realized the meaning of this sentiment until I became active in specialty seed production. Thank you ISU!

MY IVS EXPERIENCE IN VIETNAM

In the beginning of my working life, I wanted to do something out of the ordinary. This aspiration meant that I had to have a willingness to leave the familiar behind. An exceptional and unanticipated opportunity— International Voluntary Services (IVS)—made this possible. I worked with IVS for five years on agricultural projects, mainly in tribal villages in Vietnam's Central Highlands. The lessons I learned during that time were to serve me well years later when I was working with growers in Oregon. By the time I left IVS, my findings in regard to the promotion of agricultural innovations among farmers included the following:

- First, it was extremely important to establish good relationships with farmers based on mutual interest and trust in a successful resolution to some agricultural problem or opportunity. In Vietnam, I had learned many times over that, as an outsider, I was constantly being tested for my relevance and loyalty. This experience taught me to continuously focus on the maintenance of good relations with my "customers" and on the original purpose of our endeavors.

- Second, it was easier to establish such relationships with successful farmers than with subsistence farmers. Successful farmers had a deeper understanding of the agricultural fundamentals and of the risk-reward tradeoffs. They were also better-positioned to absorb losses. In addition, they were usually recognized by other farmers as leaders and could therefore subsequently act as change agents for successful innovations.

- Finally, starting small was important. Mistakes on small plots were not seriously costly and, conversely, success stories were easy to advertise.

These were basic principles of agricultural extension that had evolved in the U.S. over many years. My volunteer work in Vietnam confirmed that they applied in Vietnam as well.

GOOD RELATIONSHIPS WITH AMERICAN FARMERS

My working life in agribusiness in the USA re-confirmed my Vietnam experience many times over. Working with American farmers required the same kind of trustful personal relationships that I had had to develop in Vietnam. The only real difference in my relationships with farmers in Vietnam and in the USA is the fact that my relationships with American farmers were ultimately formalized as grower contracts.

I tried with some success to develop a stable group of growers in Oregon as Agricultural Alternatives' business expanded. We were becoming dependent on each other, but at the same time, the only thing binding a farmer to me was a contract for a single season. Our growers always had the options to switch to another seed company or to grow other crops. However, they had motivations to stay with us. They were learning production protocols and ways and means of improving the quality of their production and their expertise increased in value every year. On my side, I needed to be able to demonstrate to my market customers that I had a reliable capacity to deliver quality products. The components of seed quality included strong vigor and germination, little or no contamination, no diseases and no moisture problems. In the face of constantly changing circumstances, I was compelled to invest a good deal of my time and myself in keeping the best growers with me from year to year.

SMALL IS BEAUTIFUL

I had first learned that small is beautiful in Vietnam when I took the advice of a USAID agronomist to use metaldahyde for slugs in vegetable production in Dalat. One evening, I applied the pesticide on a small cabbage field of a cooperating farmer. The next morning dead slugs were everywhere. This event was repeated on several other farm demonstrations and I suddenly had great credibility with the local farmers.

The lessons I learned in Vietnam also applied on American farms. If you started with only an acre or two with a new crop, recovery from any mistakes was not too costly, and little productive land had been risked. For example, I tried some Japanese parsley on a couple of acres on the farm of a grower who was willing to take a small risk. We couldn't get the crop to grow well because, I suspected, the temperatures in Oregon weren't warm enough. The farmer absorbed the small loss and we both walked away a little wiser, but still on good terms.

We had a more positive experience with red radish seed. Again we started with a few growers on a few acres. We learned on these small trials that we had better results if we transplanted the seedlings. Over time, our careful trials expanded to several hundred acres of red radish seed in production.

Oregon has a lot of small growers. The average farm size in Oregon was 460 acres in 2012 according to the USDA Economic Research Service. Working small kept me in contact with small farmers and also served to keep expectations reasonable during initial trials of new seed crops.

THE HORSE AND THE CART IN THE RIGHT ORDER

In my business, the horse was finding a market; the cart was the actual seed production. Production was pointless without a buyer. Once seed buyer requirements were on the table, I could estimate margins,

Markets for any of these minor crops need to be established first; production systems come second.

Mike Chilton

determine whether we could make money by going into production, and figure out the detailed terms we could offer both growers and the buyers of our production. Often this procedure was stretched out over a couple of years as production protocols were refined with growers.

INTEGRITY IN BUSINESS RELATIONS

I have related above the episode of Ag Services blending wet vegetable seed with dry seed in an attempt to achieve the moisture standard of our Japanese customers. This effort failed and severely impacted the business relationships with our primary customers. The Japanese set high quality standards and what they perceived as attempts to deceive them were taken as personal affronts. This event ultimately led to my departure from Ag Services. In the course of this affair, I recalled a faint echo of my Dad's lectures in my childhood on the importance of having good values.

THE VALUE OF CROSS-CULTURAL EXPERIENCE

My experience with IVS in Vietnam and the Border Police in Thailand gave me a perspective on how to work with people from other cultural traditions. Thus, I was comfortable working with our international customers, particularly the Japanese, who came to comprise two-thirds of Ag Services business prior to my departure in 1985. At the risk of being repetitive, in my own business, I put special emphasis on maintaining good relations with our customers, particularly, but not exclusively, the Japanese. I paid regular visits to their home offices to pay my respects, eliciting and addressing their particular concerns.

A GLOBAL PERSPECTIVE

A knowledge of global geography is very important in the seed business. Seed production is diversified in various regions of the world to protect against crop failures or market dislocations in any given area. My international experience gave me first hand exposure to the climatic and environmental differences that affected agricultural possibilities in various places around the world.

STAYING BELOW THE RADAR

Closely related to the constant search for diversification, I found it useful to keep a low profile. Visible success in the seed business is quickly emulated. This translates into more production, reduced margins, and efforts by other seed companies to copy production systems and to poach growers. In other words, publicity reduces the life cycle of a profitable crop. Thus, I deliberately didn't share business information with the state extension service or the nearby Agriculture faculty at Oregon State University. These state institutions are in the business of disseminating information about agricultural success stories. They don't see their role as protecting the margins of agribusinessmen. I had good relations with OSU professors and staff of the extension service and occasionally sought their advice, but I didn't boast to them about any successes. In addition, we avoided ostentatious expenditures in our private life. We paid our taxes and lived modestly.

FINDING LOCAL SOLUTIONS TO TECHNICAL PROBLEMS

Over the years we became more or less self-reliant when it came to dealing with technical issues of seed production. For example, I mentioned above the problems we had with threshing daikon radish seed. A local engineer figured out how to make large rollers—like those on the top of old-fashioned washing machines—for the front end of the combines that farmers had at the time. The soft rubber rollers crushed the seedpods before they entered the combine but they also minimized bruising of the seeds. Bruised seeds had black spots on the cotyledons, first primary leaves at germination, which our Japanese customers disliked. As noted above, our engineer fabricated twenty sets of rollers that we rented to our growers for the harvesting season. We followed this procedure until the expanded availability of the more efficient rotary combines made the rollers less important.

Another example of local technical solution was developed for blueberry pro-duction. In the last few years, blueberries have expanded rapidly in the Willamette Valley. Growers found that existing self-propelled spraying equipment was too low to the ground and damaged the bushes during pesticide spraying. Rather than request modifications from the manufacturer, Stan Klopfenstein, a son of Wilbur Klopfenstein, one of my growers, treated this situation as a business opportunity and began modifying sprayers in the family farm workshop by extending the legs of the equipment so that the driver rode high above the blueberry bushes. He had already completed fourteen modifications for other growers when we interviewed him in the summer of 2014.

Establishing isolation protocols for seed production was a more complex problem. In order to maintain varietal integrity, fields need to be isolated from possible contamination from other varieties. We were able to engage the concerned seed production companies and growers and came up with a cooperative solution to the need to isolate seed fields. Isolation distances varied with the crop depending on whether it relied on wind or insects for pollination. Insect pollination could extend up to two miles for cabbage, for example. The wind pollination distance for beets was much greater—up to five miles.

Over several years we evolved a mutually agreed system that set isolation distances for each seed crop. The grower who first started a crop had seniority; that is, his neighbors couldn't grow the same crop within its isolation zone. Fortunately, seed farmers understand the importance of maintaining varietal integrity and recognized the need for such a system. If any grower in the area produced poor quality seed, it would affect the reputation of all farmers. We had occasional isolation violations, but the system was largely self-enforcing and worked smoothly. The system we set up was the result of our own efforts and it worked well because farmers cooperated with each other on the basis of mutual self-interest.

THE LIMITS OF GOVERNMENT ASSISTANCE

After residing stateside now for over forty years, I often find myself discouraged and sometimes rather cynical about the activities of government agencies and our elected leaders on the state and national levels. Since owning and operating my own company and witnessing the governmental misuse and waste of our tax dollars first hand, these feelings have gotten stronger. And yet, I must not forget that my first experiences abroad were governmentally-sponsored activities paid for by those very same taxpayer dollars. In spite of frequent discontent today, I have to look back and give thanks for the great opportunities—paid for by U.S. tax dollars—that were afforded me at an unusually early stage of life.

While I have mixed feelings about the role of government as it affected my own life, I can make a couple of comments based on first-hand experience with government assistance for agriculture. At Agricultural Alternatives, we did our best to solve our own problems without government help.

In the first place, we didn't really want government officials to get too close to our business as they were in the business of spreading the news about agricultural success stories. The faster information spread, the quicker our margins would compress.

Second, we were often reluctant to ask the nearby OSU research and extension people for help in solving our problems because their interests were more in agricultural research than in the nitty-gritty of agribusiness problem solving. We found that some state agricultural technicians were sometimes more focused on publishing research papers than in dealing with our practical problems.

Third, we didn't need financial or material support that OSU might have been in a position to provide or endorse, that is, if we happened to have an issue that was on their research agenda.

Fourth, we did find a useful role for the government agriculturalists as neutral mediators in disputes about seed contamination. So, while I had friends at OSU and maintained friendly relations with government officials, I learned that these relationships needed to be carefully managed in order to protect both public and private interests. This meant that, in some cases, I didn't share all of my business knowledge with government officials.

In any event, I had learned in Vietnam that government support for agriculture can sometimes be heavy-handed and slow-moving. Real engagement with farmers and agricultural enterprises must be personal, responsive, and useful.

THE NEXT BIG (OR SMALL) THING

With the exception of two homeruns—evening primrose for oil and radish seed for sprouts—Agricultural Alternatives was a niche player in the vegetable seed business in Oregon and Washington and that business was always in flux. To a large extent, this is due to the vagaries of the global market. Demand can shift rapidly in either direction because of weather related gluts or shortages. Furthermore, consumer preferences can change quickly and foreign exchange rates can shift for you or against you. All of these factors were beyond our control. The ups and downs of the seed markets were almost impossible to predict. When the production of a new crop

Things don't have to change the world to be important.

Steve Jobs

expanded quickly, competition eroded profit margins, making the crop less attractive to growers and to me.

My response to the unpredictability of markets and prices was to incorporate into our business model the constant search for opportunities to diversify new production. In addition, I had learned that the bottom line in our accounts (profit) was far more important than the top line (sales volume), so I was always looking for crops with better returns. Because of the constantly shifting patterns of demand and marginal returns, we often grew a particular crop for only two or three years before supply and demand moved us on to other opportunities. For example, burdock came and went; the same can be said for cilantro seed. Parsnip demand, primarily from the UK, was limited, so we grew it on an irregular basis. Annual chrysanthemum for the Japanese market was not a large acreage crop and demand for it was also irregular.

Cost Minimization

One element of keeping a low profile for Agricultural Alternatives was keeping costs down. We minimized our office overhead by working from home. My wife did our bookkeeping for a nominal salary in order to qualify her for Social Security. For other personnel needs, I contracted the professional support we needed: one field man to cover the Willamette Valley, a lawyer, and an accountant. I personally handled most of the production details in eastern and central Oregon and eastern Washington.

At one point, we did consider the merits of setting up our own seed cleaning operation. Seed conditioning operations in our region were mainly used for processing grass seed. Vegetable seed, however, has much stricter purity standards than grass seed. We initially required 99% purity in our vegetable seed contracts and over time that standard was raised to 99.5%. Meeting this specification presented a large challenge to our growers, but setting up our own seed processing facility would have presented a larger challenge to us. Different vegetable seeds required different protocols. Processing procedures were complex and regularly required fine-tuning. The management of a seed cleaning operation would have diverted a good deal of our scarce management capacity from our main task of seed production. Eventually, we identified and used a total of six seed cleaning operations around the region that were able to handle vegetable seed competently. Thus, by not branching out into seed cleaning, we were able to stay within our "light footprint" management style, keep our overhead low, and maintain a simple business model.

The Exceptional Case Of Evening Primrose

The production of evening primrose seed was exceptional in two ways: first, we grew it continuously for nearly two decades, and second, the crop was grown as a "commodity" crop for extraction of its oil from the seeds and only occasionally as seed for planting.

I had started primrose production with Ag Services with twenty-five acres each year in 1977 and 1978. When I left Ag Services in 1985, the demand for oil of evening primrose was rapidly increasing for numerous cosmetic and nutritional uses. In accordance with the terms of my non-competition agreement with Ag Services, I would not grow primrose in Oregon for a period of time, so as mentioned earlier, Agricultural Alternatives started primrose production in North Carolina and Virginia.

Efamol, our primary customer, had provided a financial incentive for us to start production. We worked with Efamol agronomist, Dr. Peter Lapinskas, to jointly develop the production protocol for primrose. After a couple of years, we recognized that primrose had better yields in Oregon and Washington so we returned to the Pacific Northwest in the late 1980s. The move was timely as demand was soon outpacing supply.

We ultimately were growing about one thousand acres in production in the early 1990s. In our best years, 1991 and 1992, we shipped about a million pounds of seed to the United Kingdom for crushing—about twenty-five 40,000-pound containers each year. Soon after this production peak, increased global demand stimulated more production in the U.S. and Europe. This increased production put downward pressure on prices and our acreage. Nevertheless, we stayed with primrose even as margins began to erode. With experience, we found the crop easier to grow. More importantly we found that we needed the regular, predictable cash flow primrose provided in contrast to the ups and downs of vegetable seed production. Primrose remained a consistent part of our production for many years.

In the early 1990s, the Chinese discovered that they could harvest wild growing primrose in China and reportedly exported 50 metric tons of seed scavenged from wild plants in their first year. Even though we produced a better quality crop, the demand for our primrose dwindled away. Later, with cheap labor and a national commitment to export development, China became the large scale, low cost producer in the global market for primrose as it remains to this day. Our last year of evening primrose production in the Pacific Northwest was 1998 when farmers in the Columbia Basin and La Grande areas grew about 250 acres.

BUILDING A PROFESSIONAL NETWORK

Agricultural Alternatives' operations dealt with the ever-changing circumstances of seed production, such as markets, prices, and unpredictable natural conditions. In this environment, we needed to be in regular contact with key players in the seed industry. In other words, we needed a friendly network of knowledgeable people and firms in our business.

To this end, we made it a point to actively participate in the annual gatherings of the seed fraternity from around the world. We attended international seedsmen meetings hosted by The International Federation of Seedsmen (IFS) headquartered in Geneva, Switzerland, and the Asian and Pacific Seedsmen's Association (APSA) based in Bangkok, Thailand. During my years in the seed business, IFS met in such exotic locales as Venice,

Buenos Aires, Brisbane, Rome, Monte Carlo, and Nairobi among others. APSA met in Jakarta, Bangkok, Bangalore, Christchurch, Manila and other cities.

These meetings proved to be well worth the investment of our time and money. The APSA events were particularly important for us as these meetings focused on the Asian seed business and included representatives of large numbers of vegetable seed companies active in the region. Though participation was expensive, these meetings afforded valuable opportunities to learn about new crops and marketing possibilities as well as to meet with buyers and suppliers. We also found that these annual events were an excellent way for Agricultural Alternatives to maintain and extend our worldwide presence in the seed industry. They also provided welcome opportunities to visit new places and make new friends.

FIVE

CROP OPPORTUNITIES

I'd like to provide some thoughts on potential crop opportunities for would-be agribusiness entrepreneurs in the Pacific Northwest. New specialty crops and marketing opportunities are going to appear with regularity. Smart agribusiness growers will watch for them and test some of the crops to keep the flexibility and diversification that success requires.

The key point to remember is that the markets for any of these minor crops need to be established first; production systems come second.

TRENDS IN SEED AND SPECIALTY CROPS

The capacities for diversified seed and specialty crop production are all in place in the Pacific Northwest: lots of entrepreneurial small growers who are willing to try new crops; near perfect agronomic and climate conditions for most cool weather seeds; a supportive public agricultural research establishment; excellent infrastructure—ports and roads for easy access to global markets; easy and efficient customs formalities. In other words, we have ideal conditions for diversified agricultural production for markets anywhere in the world and, in particular, a large potential to produce for specialty agricultural markets.

Where are the future possibilities for specialty seed and plant crops? While there are no guarantees of success and in many cases more study is necessary, here are some of the opportunities that I see:

QUINOA (*Chenopodium quinoa*)

Quinoa is a grain-like crop of South American origin grown primarily for its edible seeds. There is evidence of its successful domestication for human consumption over three thousand years ago in the Andes. The plant is of the same family as plants we commonly know in these growing regions, e.g., lambsquarter and kochia (weeds), spinach, garden beet, Swiss chard, and sugar beet. Quinoa has become increasingly popular among fanciers of organic and health foods for its nutritional value. The gluten-free grain-like food is rich in protein, a good source of fiber and high in magnesium, iron and calcium. Grown at high altitude in South America, its adaptability to other growing environments has not been well studied.

GRAIN AMARANTH (*Amaranthus* sp.)

Another crop of Latin American origin with a long history, grain amaranth has been cultivated as a food staple of the Aztecs. Like quinoa, this plant is closely related to several weedy and floral species of the amaranth family commonly growing in the Pacific Northwest. Grain amaranth is high in protein and other nutrients and also a favorite of health food consumers. Most species are quite easy to grow with production yields comparable to rice or maize. Like quinoa, the development of expanded markets for grain amaranth is currently a function of growth in the health food market.

CHIA (*Salvia hispanica*)

Yet another Latin American food crop, a species of flowering plant in the mint family is commonly known as chia. Chia seeds can be eaten raw and are a good source of protein, fats and fiber. The seeds can also be ground into a coarse flour for use in cooking and baking or as an ingredient in drinks. Chia sprouts can be used in salads or sandwiches like alfalfa spouts. And yes, the sprouts are what emerge from the pores of the once-popular clay "Chia pets". The traditional growing area for chia is in Central America. Its adaptability to our region needs to be tested by growers.

CONEFLOWER (*Echinacea purpurea*)

While the use of Echinacea is somewhat controversial, the native North American perennial flowering plant is believed to enhance the immune system of the body. It is reportedly effective as a suppressant of oncoming colds and respiratory problems. (Among other sources, this is based upon my first hand experience!). Hundreds of acres of the crop have been grown in the Pacific Northwest and processed using techniques and equipment from existing crop production systems. At this time, unfortunately, essential information about markets for this crop is inadequate for the purposes of projecting future demand.

PSYLLIUM (*Plantago ovata*)

Psyllium is the name for several plant varieties that have two main uses: as dietary fiber and for the production of mucilage. High fiber breakfast cereals have been shown to contribute to the reduction of blood cholesterol. A large proportion of psyllium husk goes into the production of Metamucil and other high fiber products. Mucilage has several uses, for example, as a thickener in ice cream and as glue.

The U.S. imports eight thousand tons of psyllium annually, mainly from India. Psyllium grows well in the cool dry weather of North Gujarat state. Psyllium fields have been planted in Washington state and Arizona. Despite the continuous increase in demand for high fiber products, the commercial feasibility of US production remains to be determined.

OTHER FIBER CROPS: FLAX, HEMP AND RAMIE

FLAX (*Linum* sp.)

Flax was an important minor crop in the Willamette Valley until just after the Second World War. With the advent of new technology, fiber of natural origins was replaced by synthetic fiber, even for sails and rope in the maritime industries that had been heavily dependent upon the resilience and toughness of flax fibers for centuries. Today, there is only limited and specialized demand for high-grade flax fibers. These include artisan uses, such as textiles, primarily in Europe, and the production of specialty papers. U.S. flax production today is concentrated on oil producing varieties for industrial uses and as a source of Omega-3 and Omega-6 oils used in health supplements. Flax is mainly grown in the upper plains states. North Dakota has 85% of U.S. production. Canada is the largest producer of flaxseed in the world. Flax production requires cool climates with long periods of daylight. Hemp (*Cannabis* sp.) has already been proven to thrive in our region! Ramie is a member of the nettle family with extremely strong fibers. Its production and marketing prospects remain to be determined.

NATURAL DYESTUFFS

Natural dyes, made from plant sources, are making a comeback after a period of long decline when synthetic dyes took over the bulk of the market for dyestuffs. The resurgence of demand for useful dyes is a direct result of growing concern about the health and environmental impact of synthetic dyes. The plant raw materials for natural dyes comprise dozens of species and are used in a large number of niche markets.

CUCURBITS (*Cucurbitaceae* family)

An important plant family that includes squashes, melons, and gourds, cucurbits are commonly found in both temperate and tropical climates around the world. This group is one of the more exciting and least studied groups of plants in regard to their potential for health and medicinal uses. For example, a plant from one genus, *Coccinia* sp., is used as a medicinal supplement for the stabilization of sugar levels among diabetics. There is also anecdotal evidence of cucurbits being used to build immunity against viral infections.

ANNUAL WORMWOOD (*Artemisia annua*)

Identified as a source of active ingredients to treat malaria in patients who have developed resistance to natural or synthetic quinine, annual wormwood has been easily grown in garden-sized plots in the Pacific Northwest. More study and analysis are needed to determine useable levels of active components that can be produced within the different climatic areas of our region.

CORIANDER (*Coriandrum sativum*)

A member of the carrot family widely used in Asia and the West as a flavoring source in the form of its processed seed or extracted oil, coriander has been grown successfully on fairly large acreages mainly in the drier regions of central Oregon. Other members of this family also commonly used as flavoring agents include celery, dill, fennel, parsley and anise. These too can also be successfully grown within the Pacific Northwest as seed crops or food crops or for processing into condiments. There is good production potential in the region for all of these crops for growers who are able to identify customers in the food, health food and medicinal markets.

ST. JOHN'S WORT (*Hypericum sp.*)

A medicinal herb used to relieve mild depression, St. John's Wort is available over-the-counter and may be prescribed by naturopaths. The plant is well adapted to the Pacific Northwest and is often found as roadside groundcover. It also can become a weedy pest as the plant is a prodigious seed producer. It is toxic if ingested by livestock. In recent years, limited acreages have been grown in central Oregon. As a commercial row crop, it is swathed at time of blossom, sun-dried and fed through a combine to separate dry leaves from the remainder of the plant.

PEPPERMINT AND SPEARMINT

Consequential minor crops of the Pacific Northwest, peppermint and spearmint are both grown for the distillation of oil. More recently some mint crops have been harvested for their leaves which are sold to herbal tea producers. This crop requires capital-intensive investment in distillation equipment whose use is required for only a few weeks out of the year. This equipment could conceivably be used with other oil crops potentially adaptable and economically sustainable under our climatic conditions.

HOPS

One of the most significant minor crops in the Willamette Valley during the earlier part of the 20th century, several fortunes with lasting legacies were made with hops. Production subsequently shifted to the Yakima Valley in Washington. The likely reason for this shift was the lower sensitivity to powdery mildew infection in the more easterly and drier desert climate in the Yakima Valley. Today, a large portion of hops used in the U.S. brewing industry comes from the Pacific Northwest, mainly Washington state. A small portion is still grown in the Willamette Valley.

Left to right, Mike Chilton, author (Turner, Oregon); Ken Wiser, Production Representative Quincy Farm Chemicals (Quincy, Washington); Tim Welsh, Managing Director, SeedAsia (Bangkok, Thailand); Bob Griffin, Co-author (Honolulu, Hawaii) pause on a tea plantation near Bogor during Asia and Pacific Seedsmen's Association conference in Jakarta, Indonesia, September 1996. Photo by Norma Griffin.

Mike Chilton makes a visit to "The Shack" near Baraboo, Wisconsin, where the renowned environmentalist, Aldo Leopold, did much of his early conservation writings. June 2004.

Mike and Simone Chilton make a stop at Columbia River Gorge observatory point, June 2002.

Top photo: Vehicle restoration has been an entertainment when not attending to seed production affairs. The above 1949 Diamond T pickup has been one such project.

Having a good dog was an important part of a home. Rex, our Border Collie, served that function for many years.

John Harding, Agricultural Alternatives Willamette Valley Production Manager, visits with Mr. Nori Oka at Takii Research Station, Kyoto, Japan.

Top photo: A well-managed weed-free evening primrose crop begins to show promise of good dividends in early spring.   As the crop comes into bloom, prospects of a good yield continue to build. Bottom photo: Early cooperators in primrose production, Jeff Boettje and Bob Hulbert, at Mt. Vernon, Washington, stand in a field coming into full bloom, 1993 season.

Top photo: Shasta daisy for seed production. Visited by Ivan Small, Agricultural Alternatives summer intern, and Jerry Newell, production representative for Central Oregon, July 1993, Madras, Oregon.

Bottom photo: A garlic chives crop grown by Wilbur Klopfenstein near Silverton, Oregon. Still in bloom on October 3, 1994, it was harvested before rains came in late October. However, artificial seed drying after harvest was necessary.

Top photo: A hybrid radish crop is harvested by Ray Palmer near Willamina, Oregon, October 1994.

Bottom photo: Seed bed is prepared for spring planting by Ray Kuenzi, Silverton, Oregon. Yellow Wallflower (*Cheiranthus* sp.) seed production in background.

Red hybrid radish in a commonly used 4:2 female: male configuration. Crops in both pictures grown by Harry Klopfenstein, Silverton, Oregon. After pollination, male plants are removed.

Daikon radish with pivot irrigation equipment in background. Commonly, pivot systems are designed for 160 acres which irrigate a circle of 120 acres, with hand set lines in four corners of about 10 acres each.

Top photo: Irregular moisture is the cause of irregular stands of evening primrose. Middle photo: Small seed is placed just beneath the surface, but all seeds do not contact moisture evenly as water travels down the ditch or 'ril' between two rows Bottom photo: Once plants have sufficiently emerged, cultivation with cutaway discs is urgent to keep ahead of developing weeds, September 1995.

Stand establishment is often difficult in specialty seed production. To assure greater uniformity, transplants are placed by machine and work crews, as we see here with a ten-row spring planting of annual brassica, Willamette Valley, Oregon.

Hybrid red radish in full bloom growing near Willamina, Oregon, is inspected by John Harding, Agricultural Alternatives Willamette Valley field representative, July 1994.

Top photo: General view of adzuki bean seed production field inspected by Don Panek, Agricultural Alternatives computer specialist, who was enjoying one of his many up-country inspection trips. Bottom photo: Adzuki beans grown for the Japanese market have set pods in Columbia Basin production, August 1994.

Photo left: Somewhat irregular flowering and occasional off-types (pictured here, height difference) add to the difficulties of high quality onion seed production, June 1994.

Photo right: Off-type bulbing onion plants are identified and eliminated by Brian Davis, field representative for Quincy Farm Chemicals, Quincy, WA.

Irrigation equipment is essential for most specialty seed production. Top photo: Hybrid spring *brassica*—2:2 ratio, complete combined harvest. Grown by Randy Rohde, Amity, OR. Bottom photo: A red radish hybrid—2:4 ratio—in which males are removed after polination and only females are harvested. Grown by Don Fisher, Junction City, OR, July 1993.

Top photo: 5,000 plants-per-acre is ideal for burdock. Transplanted to the field in June, burdock is harvested one year later in September by conventional combine.

Bottom photo: Biennial burdock is fun to grow but seed harvesting is miserable because of irritation by the very small, unlimited number of 'flying hooks' which break away from the burrs of mature plants during harvesting.

'Ril' irrigation—every second row—for bulbing onions is attended by Brian Davis of Quincy Farm Chemicals in the Columbia Basin, Washington.

An excellent stand of bulbing onions from direct seeding in the Columbia Basin, Washington, April 1994.

Because of small seed size, evening primrose optimum stand establishment is most difficult to obtain. Farmers in the Columbia Basin, Washington, have developed a technique of cross cultivation for thinning. After much trial and error the technique has been perfected.

Top photo: Evidence of thinning field by cross-cultivation.

Bottom photo: Field after cultivation. John Biersner, Columbia Basin Production Representative, gives thanks for a well-established primrose stand,1994.

Seed from the field is brought in bulk to warehouses where it is unloaded into bins to await processing (cleaning). Here, radish seed is unloaded at B & M Seed Cleaning in Salem, Oregon, by owner, Dick McGuren, and his staff, September 1994.

SIX

INFLUENCES THAT PREPARED ME
FOR ENTREPRENEURSHIP

*B*y 1983, *the vegetable seed production tail* was wagging the Agricultural Services dog, at least in terms of the firm's profitability. Contract grass seed production and blended fertilizer sales were low profit margin businesses. The margins on daikon seed were riskier but much stronger. At that time the specialty seed operation had about thirty-five hundred acres of various vegetable seeds in production: twenty-five hundred acres of daikon and one thousand acres of cabbage, peas, parsley and other vegetable seeds. Ag Services had become a major provider of daikon seed to the Japanese market. Half of the firm's staff was working on vegetable seed production and processing.

THE TURNING POINT: WHY I STARTED MY OWN BUSINESS

I had also become increasingly discontented with the lack of remuneration to me for the success of the vegetable seed business and started to agitate for some kind of commission over and above my salary. At first, the owners were unmoved, but eventually they came up with about half of what I had asked for.

In the 1984 growing season, the demand for daikon seed from Ag

Services' customers dropped significantly. This decline was partly due to the seed quality issues that the company had created for itself the previous year and partly due to an overall saturation of the market. The period of explosive growth in the daikon seed business had ended. The pattern that I would see repeatedly in the coming years became clear: good prices attracted more growers to produce a crop that created an over-supply of the commodity and subsequently lower prices. I needed to focus my energies on other opportunities.

Company issues came to a head late in the year. At the peak of the Ag Services' profitability and sales, the owners had managed to secure a buyer for the company. I was leaving for a post-harvest trip to Japan when the owners revealed their decision to merge with a firm named Native Plants Incorporated (NPI) of Salt Lake City. NPI was basically a seed research and breeding firm. NPI's owners wanted to add a production arm to provide cash flow to support its research operations and Ag Services fit the bill. After returning from Japan, I visited NPI and failed to see any synergies with Ag Services' operations. I could see that this new arrangement offered no future for me, and I decided to leave Ag Services on April 1, 1985, and develop my own seed business.

I started as a sole proprietor. As my seed business grew, I set up my own company, Agricultural Alternatives Ltd., to take advantage of a corporate structure for my operations.

In retrospect there was probably more push for me to leave Ag Services than pull to set up my own company. But was it courage or foolhardiness, confidence or over-confidence, foresight or denial of reality that led me to taking on the risks of self-employment? While my dim view of the future working with Ag Services' new parent company may have triggered my decision, my entire body of life and work experience and the lessons learned came into play. I will devote the remainder of this chapter to presenting brief descriptions of these critical influences on my life, influences that sent me down the road to entrepreneurship.

Rural residence on farm near Sigourney, IA, where the Chilton family was raised. Photo, circa 1960.

GROWING UP ON AN IOWA FARM

I was born towards the end of the Great Depression in Iowa City, Iowa, where my parents were students at the University of Iowa. My father, Ralph, was originally from Connecticut. His biological father died when he was three years old and his mother passed away when he was seventeen. For a while he lived with his stepfather and attended the University of Connecticut. Soon, the famous advice of Horace Greeley somehow affected him and he went west, stopping at Iowa City where he enrolled at UI. There he met and married my mother, Verna Long.

Despite their interest and strong belief in the value of higher education, neither of my parents finished university. Grandfather Long purchased an eighty-acre farm in 1939 for $50 an acre to help them get established and have a place to live. The farm was near Sigourney, Iowa, my mother's hometown. Sigourney was founded in 1844 as the county seat of Keokuk County. The town's population was 2,355 in 1940. By the time of the 2010 census, it had fallen to 2,059.

Mike Chilton with his parents, Ralph and Verna, during a light moment while travelling in northern Italy, fall of 1968.

Ralph and Verna Chilton had six children of whom I was the eldest. My five younger siblings were born into the family at intervals of two to three years. After me, in order of birth, were three girls, Patricia Ann, Natalie Jane, and Sarah Jo, and two boys, Anthony Thomas and Donald Allen.

Patricia Ann (Pat)

> Pat grew up to take a nursing degree from the University of Iowa and subsequently had a full career in nursing. Pat married John Malone, a forester, and raised three boys. She is now retired and living in Wenatchee, Washington.

Natalie Jane (Nat)

> Nat also graduated from the University of Iowa. She studied Italian literature in Florence, Italy and returned to the USA to work on a PhD at Cornell. Eventually Nat changed directions and became a real estate investor and manager in Buffalo, New York. She now lives in Mexico.

The six Chilton siblings on retreat at Garibaldi Beach on the Oregon Coast, July 2004. Left to right, Don, youngest brother, and his wife, Judi; Mike and Simone; first sister, Patricia, and her husband, John Malone; third sister, Sarah, and her husband, Greg Stiegmann; Pam, wife of younger brother, Tony; and second sister, Natalie.

Sarah Jo (Red)

Red graduated pre-med from the University of Iowa and became a physician. She married a fellow medical student and specialized in radiology. She raised two children and after working in Canada and Colorado, she is now retired in Denver.

Anthony Thomas (Tony)

Tony attended William Penn College in Oskaloosa, Iowa, before heading to the Pacific Northwest, first for summer jobs and then full-time work mostly with the U.S. Forest Service. He worked with his brother-in-law, John Malone. For most of his adult life, he specialized in heavy equipment operation and maintenance, mainly large mobile cranes. He married a Northwest girl, Pamela, from Mount Vernon, Washington. They are now settled in Longview, Washington, where they have raised three children.

Donald Allen (Don)

After spending some time working with Tony on forestry projects, Don became the fourth family member to take a degree from the University of Iowa. He majored in Sociology and after graduation, he moved to the Pacific Northwest to work with county Social Services in Longview, Washington. He married and had two girls. Subsequently, he worked with the Bonneville utilities program on energy conservation and finally with Freightliner Manufacturing before moving and retiring in the Columbia River Gorge area near Lyle, Washington.

We were all raised on the farm and were steeped in farm life. The farm was a subsistence operation that produced grain and forage crops for our livestock—cattle, sheep, pigs and chickens—that we sold locally. The farm barely broke even. "We couldn't afford to stay, but couldn't afford to leave," Dad used to say. All six of us contributed our labor to enable the farm to support us.

One episode from these early years stands out today as I look back. It was my first exposure to the vagaries of farm economics. I was around ten years old when I realized that geese were commanding quite a high price in the local market. A neighbor of my grandfather had a lovely pond where he had kept and raised a number of geese over the years. I had developed a deep interest for birds in general and would always admire this flock of snow-white geese whenever we would pass by on the way to our grandparents. I thought about the grand opportunity of raising birds for high market value. One day, I approached the owner to see if he would ever be interested in selling his flock. He replied with an encouragingly affirmative "yes"! That fall, with money earned from lawn mowing in the neighborhood, I purchased and moved all eighteen birds (six ganders and twelve females) to our home. I built pens, feeders and a basic shelter to protect the birds from the bitter winter cold.

As egg production commenced in late February, I had figured out how the eggs could be hatched under poultry hens. We had available two nesting batteries totaling twenty-eight cubicles, which I filled with non-laying broody hens purchased from the laying flocks of nearby neighbors. Setting six goose eggs under each hen for the twenty-eight-day incubation duration, I was soon in the demanding business of managing a gosling hatchery! As the goslings emerged, they had to be placed under heat bulbs, fed and watched over with great care. The new goslings were of differing ages and numbers because of a growing supply of eggs from the parent flock. By May, I had hatched approximately 150 goslings and was in need of space. As young birds became less dependent upon heat and began to develop primary wing and body feathers, I would place them into a recently fenced off area of young plants and weeds. They soon became accomplished grazers.

Coming into summer, numbers increased and the goose project was moving along well, until one day I noticed something ominous: some of the youngsters were becoming droopy and soon dying. Feed, care, water, the compound, aerial and other predators all were re-evaluated to determine what might be the cause, but the young geese continued to die. By the end of the summer, I had lost about half of the total hatch. To this day, I am without a final answer for that heavy mortality, but I suspect that within the diverse spring vegetation that they fed upon there was a plant which proved toxic to the young birds. I finished the season with about seventy matured birds ready for market that fall. I presumed that they would be the main courses of many local Thanksgiving and Christmas dinners. When I was ready to sell the birds, I found that premium price for geese of earlier years had collapsed. I sold the young birds to a local buyer, who was a personal family friend. With the final proceeds for the year's effort, after accounting for costs, I barely broke even.

In hindsight, this was my first personal experience of the unpredictability of agricultural production and marketing systems over which the producer has little control. Later in life, when I became an entrepreneur in specialty seed production, the importance of establishing and agreeing to market prices at the beginning of a production cycle was a lesson I didn't have to learn.

Dad had his own approaches to making ends meet for our family. He resorted to a variety of part time jobs and moneymaking activities. He worked for almost a year on the Alcan Highway and later he spent some time in Greenland with the crew building facilities for the Early Warning Strategic Defense System. Between Dad and local tenants, the farm produced a crop each year, but Dad's real interests lay with the animals - Duroc hogs, Corriedale sheep, poultry, and both milking and beef animals. He became quite interested in cattle, mainly Milking Shorthorns, which he bought and sold throughout the Midwest. In yet another sideline, he organized his own transport service for Angus cattle purchased at high-end auctions around the country.

The rural life of my childhood was essentially an exercise in getting-along with what was available, a continuing pattern of tending livestock and assuring crops were properly managed in order to provide the basic essentials of life. In other words, our family did what we had to do, as did all the neighbors around us. There was little in the way of entertainment, fortunately, as we had little extra cash. As a result of shared farm work and our somewhat isolated situation, we became and remain to this day a close-knit family. In addition to our immediate family, my mother had umpteen relatives in the Sigourney area. Picnics and other events that included the extended family often drew a crowd of eighty to ninety.

I asked my brothers and sisters to give me their own thoughts on the advantages and disadvantages of having grown up on the farm. Here is a sampling of their reflections:

Pat: "I think our experiences and opportunities created a good
        work ethic and were wonderful life lessons. In those days,
        most farm kids were expected to do their fair share. One
        learned that you "reap what you sow". A farm, especially
        sixty-five years ago, afforded the opportunity to understand
        what it takes to put food on the table, referring to growing
        the meat that we ate, the dairy products, the vegetables that
        we ate fresh or preserved, and the fruit that we canned or
        froze. We learned it could be very satisfying to have a weed
        free garden, a clean barn or chicken coop, or a basement full
        of canned goods. (Just ask us. We were expected to put in
        two hours per day regardless of the heat.) We also learned
        that our friends from town found it much more exciting to
        do those things than we ever did! We worked hard as kids
        growing up, but didn't particularly question it at the time.
        We learned how to work and developed many skills in

the process. We also learned in a very natural way about science and nature, the birthing of animals and fowl, from beginning to end, and what it took to make a seed grow.

At nine years old when we finally got electricity, I learned how wonderful it was to turn on a switch to produce light from something other than a kerosene lamp. Now we could have refrigerated desserts, as well as an electric vacuum to assist in cleaning the house.

We did not grow up with silver spoons in our mouths. We knew we didn't need to be coddled to exist in the world!"

Sarah: "I realized through the hard work of survival on our farm that I would not stay on the farm in my future life as I knew it then. The farm gave me a phenomenal appreciation of Mother Nature's beauty, education in astronomy, botany, zoology, both domestic and wild. It taught me the essence of survival in every corner of life, how life is propagated in the animal and plant kingdoms, how food products are grown and ultimately reach the table. At the time, one was so busy getting through the day that such contemplation did not exist. All that occurred after leaving the farm and maturing."

Tony: "Hard work does pay off in the long run. We were brought up doing most things the hard way!"

Don: "Having been raised on a farm, my foundation stones for dealing with life were much more firmly planted. My perspectives were broader than had I been raised in the city. I have a greater appreciation of the simpler things in life now."

Although he did attend the University of Iowa, my father was largely self-educated. He was a voracious reader. His mental processes generally worked differently than a typical resident of Sigourney. Dad often talked about the importance of having good values. At the time, I wasn't quite sure what he meant, but I came to understand that he was talking about accountability, honesty, truth, responsibility, etc. His message was that life is much easier if you face life the way it is—not the way you want it to be—and stick to the truth.

With Dad often travelling on one or more of his various income generation schemes, the day-to-day management of the farm and the family fell to our mother. It was up to her to keep the place in order, the animals tended to, the kids fed and in school. As Pat mentioned, we all were required to work two hours per day at farm chores. We all had our own assigned chores, none of which could be considered fun. Mother would assure that our assignments were completed before we departed for school.

Mother grew up as the only daughter of a fairly prosperous farm family. Her early childhood was rather gilded and protected. Later, as she became a wife and mother, out parents relocated to the very basic facilities of the 80-acre farm only three miles from where she grew up. She was constantly reminded of the contrasting circumstances and new realities of her life. I'm sure it was not easy for her, but she met these new challenges directly. She provided the basic organization, discipline and leadership to meet our growing family's requirements of healthy food, clean clothes and sheets, mending, house cleaning, school assignments and attendance, doctor's appointments, bills paid, lawn mowed, gardens weeded, fruit picked, etc. She was a complete homemaker making do with extremely modest means. She was very frugal and had a good economic sense of our household budget. This was indeed one of her strong points in stark contrast with the free spending of our father. The extreme differences in our parents' approach to the use of family financial resources were a source of continuing unpleasant confrontation.

Again, my siblings had their own takes on how our parents influenced our lives and the values they tried to instill:

Pat: Our parents taught us how to work together, how to be responsible and independent, a work ethic, to respect older people, and the value of an education. The values they emphasized were "honesty, independence, self-discipline, tolerance of differences in other individuals, respect, and my mother taught us how to be frugal. Somehow we (I) learned the value and importance of family, not only our immediate family but also our extended family members on my mother's side who lived nearby. I don't remember my parents trying to impart wisdom or really engaging me in conversation a lot. What I learned was through observation of them being role models.

Sarah: Our parents were very clever in letting us know that the farm was not our life's commitment, but rather that higher education was there to be sought, and was expected of all of us, although we children knew that we would have to take care of it ourselves financially. We were always commended for good grades, occasionally got pennies as rewards, never penalized for poor grades (which we never got really). The values instilled were that of focus and hard work. We didn't have philosophical discussions.

Dad was an avid reader, and while I really was enamored with that possibility, we never had time for it. I knew that we were smarter than a lot of our peers, but that was never discussed in a condescending manner. Community was never really discussed, but both parents always held our door open to the social desires and needs of others. Everyone and anyone was always welcome to our home, whether we ate gruel or a very fine meal. Our parents taught us the art of being frugal, accepting that which was handed to us, self-discipline, tolerance of others, and much more. However, farm life created many hungers in my soul to overcome once I left the nest, which we siblings all did very readily and early in life.

Tony: Our parents taught us to think for ourselves, not to be afraid to go outside the box. Our parents imparted that we are all individuals and that even though some of our thoughts won't be down the center of the track, we should stand by those ideas.

Don: They taught us responsibilities and had expectations for each of us. They stressed higher education as one way of reaching for what was beyond the farm. Each parent was coming from different angles on our upbringing but each had their strengths. Dad was a visionary and intellectual, whereas our mother was

a realist, living here and now, and hands on. They each taught
us that we had a responsibility to ourselves and for our futures.
They stressed honesty and hard work.

There is a pretty clear consensus among the Chilton children that we
all understood the bright line that connected good grades, higher education,
leaving the farm and a better or easier life, at least one that did not require
hoeing endless rows of weeds or getting swatted in the head by a cow's
swinging tail while you were milking her. We knew we had to move on
somehow, but we also knew that we would have to manage that process
with our own resources and ingenuity.

One source of off-farm enjoyment in which I indulged was music.
Because I attended a country school, it was not until my high school freshman
year that I was able to develop any musical skills. Merrill Brown, the band
director, fortunately took an interest in me. Through him I was able to
gain a position in the bass horn section where Sousaphone players were
urgently needed, although my dream had been to play snare drums. I had
to learn fast but was able to immediately be a part of marching and concert
bands, ensemble groups, and vocal presentations as well. These experiences
became quite rewarding in later years, as the effort had given me a greater
understanding and appreciation of music, which I would otherwise not have
had. Even though fulfillment of one's dreams might not be possible, the
experience taught me that good could also come from pursuing a second
best option—joining the base horn section!

Of greater substantive significance in high school, I also had the good
fortune to be taught science by Harrison Seip, an inspirational teacher who
instilled an academic interest in me in science and math. In addition to my
Dad, Mr. Seip engendered a desire in me to broaden my horizons beyond the
family farm. As a boy, I did my share of the chores and raised my own cattle
and hogs. Our neighbors were farmers and I worked for wages on their
farms to earn and to learn. I saved my hard-earned money and followed
the college prep curriculum in high school. In 1954, I was a member of the
graduating class of fifty-two students at Sigourney High School.

IOWA STATE COLLEGE

By the time I enrolled at Iowa State College (ISC), I had acquired a good

Haig Kapooshian (from Lebanon) and Mike Chilton as grad students at Iowa State University, 1959.

deal of knowledge about Iowa agriculture and a strong sense of self-reliance. I was eager to discover the "outside world", i.e., the university and Ames, Iowa.

I entered Iowa State as the school was coming of age. I took my Bachelor's at Iowa State College and by the time I had completed my Master's, it was Iowa State University. (Pat reminded me that four of the Chilton farm kids graduated from the "more prestigious" University of Iowa and only one, Michael, from the ag school, Iowa State University.)

In college, I continued the pattern of frugality and working to support myself that I had learned on the family farm in Sigourney. My immediate problem when I arrived on campus was how to pay for the cost of living. I had a scholarship for tuition for my freshman year but I needed a job to pay for food and living expenses. I found one with the food services of Memorial Union, the college's student union. I roomed off-campus for seven dollars a week—which was considerably cheaper than the dormitories or fraternities.

In the summer between freshman and sophomore years, I shipped off to Middletown, Ohio, to work for the Armco Steel Rolling Mills in its research lab. One summer in high school I had worked on the farm of a large local landowner, Dr. Anson Hayes, who had previously done research on the development of stainless steel. He liked how I worked and he recommended me for a summer intern position at Armco.

The following summer I worked construction for a firm that was building a dormitory at Parsons College in Fairfield, Iowa. The next summer I drove a truck, our 1948 Diamond T, for my father who delivered high-value Black Angus cattle from cattle auctions to wealthy ranchers who had bought them. In the course of his self-education, Dad had made a hobby of studying cattle lineages. He regularly attended Black Angus auctions where he chatted up buyers and offered delivery services for their new purchases. One of his customers was Dr. Armand Hammer, the founder and CEO of Occidental Petroleum. That summer, Dad was injured in a traffic accident and couldn't drive so I took his place behind the wheel. Including trips to Dr. Hammer's farm in Red Bank, New Jersey, and wealthy buyers in North and South Carolina and Oklahoma, I drove fifteen thousand miles in six weeks to deliver their cattle. This experience brought reality to the romance of driving a truck to see the world. I was highly motivated to return to college that Fall.

During the early days of the 'cold war', Russian Premier Nikita Khruschev made a much publicized visit to Iowa State University before he traveled west to Coon Rapids, IA, to visit old friends at the Garst and Thomas Seed Co. Here, under heavy security, he waves to the crowds from behind the windows of the Home Economics Building on campus.

In my sophomore year at Iowa State, I had second thoughts about majoring in ceramic engineering, a field I had thought held the promise of future technology and highly paid employment. I had thought that ceramics would be a critical support technology for nuclear power. Alas, the course work was uninspiring. The professor announced on the first day of a beginning class in mining techniques that he expected his students to create a notebook "just like his" as the product of the term of study. In other words, he would talk and we would copy. I completed the class but was very disappointed with rote learning. After an encouraging visit to the Botany Department, I switched my major to botany. The botany major was considerably broader in its educational requirements than ceramic engineering, including non-technical courses such as foreign language, social and natural science, history and with out-of-class studies in photography, rock hounding and rock polishing. The Department also had a congenial faculty. While my engineering advisor worried that changing majors indicated that I was drifting without purpose, in fact I was figuring out what I didn't want to study. Botany proved to be a good deal more satisfying (and fun) than engineering. I received my Bachelor's degree in General Science in 1958—then what to do?

At that time, most botany majors who completed a Bachelor's degree became teachers. However, teaching didn't appeal to me. One of my botany professors, Duane Isely, took an interest in me and encouraged me to continue

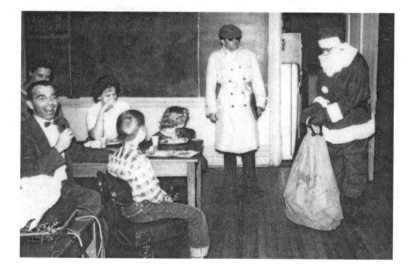

Last Christmas party before departing for SE Asia. Mike Chilton (detective) monitoring Santa (Arnold Larsen) of Botany Dept., Iowa State University. Ames, IA. Dr. Duane Isely, Chilton's major professor at left, 1959.

botany studies in graduate school. He helped to arrange an assistantship for me, funded by the U.S. Department of Agriculture, which included the princely stipend of $150 per month. I was expected to carry out research on "seed germination under different moisture conditions," which became my thesis topic. Professor Isley set up an excellent and diversified curriculum in plant sciences for my graduate study. The course work was considered general for graduate studies, but there was an emphasis on seed technology and economic botany. I finished my Master's degree in July 1960.

THE LURE OF THE EXOTIC:
GOING OVERSEAS WITH INTERNATIONAL VOLUNTARY SERVICES (IVS)

Though it ultimately proved quite useful, fulfilling the Master's degree requirements was a way of delaying a career decision. I had eliminated teaching as a possible vocation but as I approached the completion of the Master's program, I wasn't any closer to deciding what I wanted to do next.

A casual conversation towards the end of the school year changed all that. I had been teaching a course in weed identification to undergrad botany and agriculture students. One day after class, a student named Dennis Strayer asked me what I intended to do following graduation. I still didn't have an answer for that question and so I turned it back to him: what was he going to do? Dennis told me that he was considering following the footsteps of his cousin, a recent ISU graduate with a degree in Electrical Engineering. His cousin was working as a volunteer in Laos developing a speech clinic. Dennis went on to explain that his cousin's employer was a contractor funded by the U.S. government's economic aid program for Laos. His cousin's assignment entailed a two-year commitment and, though the position was voluntary, a living allowance, housing and money for basic needs were provided. The name of the contractor was International Voluntary Services (IVS), which I subsequently learned, was established in 1953 as a non-sectarian volunteer service organization. IVS ultimately served as a template for the Peace Corps, a volunteer program that was created by President Kennedy at the beginning of his administration.

At the time, I couldn't tell you where Laos was on a map, but here indeed was an intriguing situation: Laos! Speech clinic! Electrical engineering!

Volunteering!? The idea of practical work in an exotic location immediately appealed to me, especially in contrast to a career as a seed laboratory technician in Idaho—the research scientist that my professors had in mind for me—or the teaching profession my classmates anticipated. I sent off a letter of inquiry to IVS's Washington headquarters and soon received an encouraging response. There were no openings in Laos for an agriculturalist but would I be interested in Vietnam? I hastened to determine the whereabouts of Vietnam and with some difficulty was able to locate it. (Most maps, at that time, still referred to Laos, Cambodia and Vietnam as French Indochina.) Well...Vietnam, why not?

Do you want to know who you are? Don't ask. Act!
Action will delineate and define you.

Thomas Jefferson

I really didn't know any less about Vietnam than I did about Laos and I had inquired about Laos only because Dennis's cousin was already there!

Perhaps not surprisingly, the possibilities of an overseas assignment did not excite either my ISU professors or classmates. Why would I take a position without monetary remuneration, asked my fellow students? How could I consider giving up my "seniority" in seed science, my advisor wanted to know? How could taking two years out of one's life to live in a faraway place be an appropriate way to begin one's professional career? However, two years of graduate school and the responsibilities of my assistantship had already put me near the end of my endurance with academic life. The exciting prospect of working in another part of the world gave me the extra strength that I would need to complete my graduate school obligations.

To the extent they thought about it, my siblings were bemused but supportive of my decision to head for Vietnam. Sarah had to look up Vietnam on a map (as I did) and though she had never heard of IVS, she was excited about the idea of going off to the "far corners of the earth" and proud of my decision. Pat was impressed by the opportunities that IVS presented, but couldn't understand why I would want to go to Southeast Asia.

Eventually, I received a formal letter of acceptance from IVS for a horticultural position in Vietnam. My ISU graduation ceremony was in July 1960 and I departed for Southeast Asia a few days afterwards. At twenty-two years of age, I had a newly minted graduate degree, no lingering amours, and no responsibilities to anyone. As a consequence of the generous assistantship, I was also debt free. I was on my way, I hoped, to a great adventure!

SOUTH VIETNAM IN 1960

In hindsight, I can see that my arrival in South Vietnam in 1960 came at a particularly auspicious time. The Geneva Accords of 1954 had set the terms for the peace agreement that brought about an end to the First Indochina War (1946-1954). France, which had been fighting to retain its colonial possessions in Indochina, agreed to withdraw its troops from the region. French Indochina was now illustrated on maps as three countries: Laos, Cambodia and Vietnam. Vietnam was to be temporarily divided along the 17th Parallel into North and South Vietnam until elections could be held to unite the country.

The proposed elections were never held. In the six years since the Geneva Accords, a new but fragile government headed by President Ngo Dien Diem had been set up in the south. The Diem administration had many similarities with French colonial rule, but it had shown its mettle by eliminating the Binh Xuyen, the powerful criminal organization with control over many sections of the country. The former French colonial officials had departed and there were only small numbers of Americans in-country. At that time, America was a curiosity to the Vietnamese. America's life-styles, technology, movies, music, language were not well understood, but watched and emulated. As a result, my IVS colleagues and I were often the focus of that curiosity. In addition, many Vietnamese were gracious towards Americans. Perhaps they were aware of the U.S. government's lack of enthusiasm for the restoration of the French and British colonial empires. In any event, the Vietnamese made an extra effort to allow us to become a part of their lives and understand their society better.

After I arrived in Saigon, I was briefed on IVS activities in Vietnam for a couple of weeks. I was then assigned to the horticultural experiment station in Dalat, the beautiful hill station north of Saigon where Vietnamese of means would escape the hot season for cool mountain air. I joined an

IVS agricultural team whose members were posted at experiment stations located in rural areas throughout the country. The team members would meet in Saigon every few weeks to share experience and meet with Ministry of Agriculture officials. The Vietnamese we worked with really didn't understand the volunteer aspects of our role and were convinced that young men working individually on up-country assignments must be CIA agents. (IVS had a firm policy in place that prevented volunteers from compromising their service by engaging in intelligence activities.)

Working For IVS With The Vietnamese People

At Dalat, I worked alone for a while, learning the language and becoming familiar with my new work environment. I was living at the IVS house, a gathering point for IVSers and U.S. Overseas Mission (USOM, the predecessor of USAID) personnel who enjoyed stopovers and weekends in the cool and scenic wonder of the Vietnamese highlands. The villa was an elegant holdover from the French colonial period. The large house came with a four-person staff of cook, maid, interpreter and driver, whose oversight and support were added to my agricultural responsibilities at the experiment station. Needless to say, this new role in "international personnel management" was an unexpected task for a new volunteer in Vietnam, formerly an Iowa farmer.

My job at the experiment station went far beyond my expectations. The Vietnamese station manager, a recent graduate of Bao Loc Secondary Agricultural School, seemed little schooled in management. His purposes and direction were unclear, other than keeping good rapport with staff and workers and the gates open. Technical direction was forthcoming from a Taiwanese advisor from the Joint Commission on Rural Reconstruction (JCRR). The JCRR was a joint program of the U.S. and Taiwanese governments that played a critical role in the successful agricultural development of post-war Taiwan. Mr. Chuong Hsiung Yu, who was assigned to the station shortly after I arrived, was an excellent horticulturalist with many good ideas for Vietnam.

I soon realized that my years of botany studies were only modestly useful in Vietnam. The agricultural practices, the environmental factors and the overall farming environment were simply very different from my training and experience.   Furthermore, there were no readily available supporting materials on the local plants, at least not in English.   In contrast to our eighty-acre farm in Iowa, Vietnamese farmers were working two to three acre plots entirely by hand with hoes and shovels.   Their crops varied by agronomic region.   Vietnamese farmers around Dalat grew vegetables for the Saigon market; tribal farmers in the Central Highlands— called "montagnards" in accordance with French colonial usage—grew upland rice.   Although the French had kept Vietnamese farmers out of the Central Highlands, the Vietnamese government was relocating them there to establish their presence.   To do so, the govenment had subdivided some of the old French rubber plantations into small holdings and cleared land for new rubber plantations. While the Vietnamese government saw rubber as an important foreign exchange earner for Vietnam on the international market, the Vietamese farmers, however, some of them learning for the first time how to make a living from tapping rubber trees, saw the near-term production of annual crops as much more important for their survival needs.   In sum, I recognized that I had a lot of learning to do so I kept my eyes and ears open and my mouth shut in order to benefit from Mr. Yu's professional expertise as an Asian horticulturalist. He took on the task of organizing the experimental work of the station. I concentrated on learning local agriculture from him and from the farmers and gardeners of the Dalat area.

BECOMING AN AGRICULTURAL EXTENSION AGENT

Initially, the newness of my situation was overwhelming: the people, the language, the tropical climate.   I was soon in awe of Vietnam and the Vietnamese.   The place was so different from anywhere in my experience. The people were friendly and pleasant, but they lived by different values. Relationships, especially with government agriculture officials, were far more formal than the easy relationships I had had with my professors in college.

I noticed early on the contrast between Vietnamese people who lived in urban versus rural areas.   City dwellers were much more sophisticated. They were far more likely to speak French and to have family ties to France.

ie

8/27/19, 7:57 PM

Almost all rural dwellers were subsistence farmers, scratching out their livings on two to three acre plots. The hierarchical structure of families was common to all Vietnamese, however. Fathers were the ultimate authority figures, but mothers were instrumental in family decision-making. The formal family structure was literally foreign to me. Boys were pampered and spoiled. Girls, regarded as second-class family members, became diligent and responsible. It took me quite a while to absorb this new environment.

Over time, I came to realize that as a volunteer, I was an extremely insignificant entity. To have any impact at all on Vietnamese agriculture, I had to leverage my activities by making the right choices. Eventually, I learned that working quietly and on a small scale, with potentially successful ideas and with the right people, was a strategy that could yield positive results. Finding a workable solution to a known problem at the farm level tended to create its own momentum for broader application. In other words, while Mr. Yu was doing research at the station, I was becoming an agricultural extension agent.

The commercial vegetable growers around Dalat were most amenable to this approach. They were already successfully producing butterhead lettuce, carrots and cabbage. We introduced a number of new vegetable crops and production techniques to these farmers. Successful examples included bulbing onions, Irish potatoes, new varieties of cabbage, and slug control. Bulbing onions were spectacular and easily grown from the beginning, but finding markets for them took some time. Irish potatoes did well in the Highlands and became a favorite especially with the growing expatriate market. New varieties of cabbage offered insights into flavor, shelf life, and other production characteristics that were more favorable than the traditionally long-keeping, but fibrous and not very tasty local varieties. As reported earlier, in the case of slug control within the ubiquitous cabbage gardens, the use of an imported aromatic bait (metaldahyde) proved to be extremely successful. Trial results produced highly visible results: dead slugs everywhere. This successful innovation built goodwill among growers who became much more interested in our proposals on other issues in vegetable production and marketing.

By starting small, accessing Mr. Yu's expert advice, testing solutions to agricultural problems, building trust among farmers, and demonstrating real solutions to farm problems, I had become an agricultural extension agent on the job.

LEADERSHIP DEVELOPMENT: BECOMING AN IVS TEAM LEADER

After my first year in Dalat, I was transferred to Saigon where I assumed team leadership for volunteers located in the highland provinces. While still dealing with agricultural and horticultural issues at the experiment stations, I was also given responsibility for a new project for tribal people of the highlands who were being relocated due to decreasing security in their remote locations. This project, the "Montagnard Development & Training Centers", would establish centers in eight provinces to provide educational, agricultural and health services at the village level.

My team leadership role continued into a second tour with IVS from 1963 to 1965. IVS was growing as American efforts expanded in Vietnam. My new responsibilities covered supervision of a five-man Malaria Control team whose work included operation of field laboratories, providing educational extension, and vehicle maintenance. I was also tasked with supporting the administration of a "Rural Youth Program" within the College of Agriculture in Saigon, where agricultural students were assigned to weekend, summer and special projects within villages in order to learn more about conditions in rural Vietnam.

Serving as a team leader required a set of skills that was far different from those of a hands-on volunteer horticulturalist. In order to support team members in the field who were developing hamlet and village projects, I had to learn how to plan and organize work, how logistical systems functioned, how to communicate over time and space, and how to ensure accountability for our activities In addition, we all had to learn how best to work in a complicated organizational environment. On one hand, we were working through and in support of

> *The right solution to a known problem tended to gain its own momentum.*
>
> Mike Chilton, *The Fortunate Few**

* Thierry J. Sagnier, *The Fortunate Few: IVS Volunteers From Asia To The Andes*, (Oregon, 2015), 94.

the local Vietnamese governmental apparatus, but at the same time we were also an operational component of the U.S. aid program for Vietnam. Learning how to navigate these two bureaucratic systems was often more frustrating and time consuming than the work we were supposed to be doing. Despite the complexity of our institutional working environment, we grew to understand that Vietnamese interest and support were critical to our mutual success.

The "Montagnard Development and Training Centers" directed attention to ethnic groups who had received little attention from the Government prior to the initiation of the program. Subsequently, the Ministry of Education took over the program and incorporated it into its regular activities. Both the College of Agriculture's "Rural Youth Program" and the USAID-sponsored "Summer Youth Program", for which I later served as project manager, focused on developing future leadership. Thanks to these programs, several leaders and participants had been provided opportunities for further schooling abroad and later assumed leadership roles in Vietnam and United States.

> *The Vietnamese were friendly, patient and gracious...*
> *I eventually realized that the values and wants of most were not too different from those I had learned growing up.*
>
> Larry Laverentz, *The Fortunate Few*

In sum, my years as an IVS Team Leader enabled me to acquire essential management skills through self-directed on-the-job training and a much broader understanding of the developmental and security needs of the rural, particularly tribal, communities. After I completed five years with IVS in Vietnam, I was enamored with life in Southeast Asia. I set out to search for other opportunities where my Vietnamese work experience might be a useful background.

LEARNING MY PHYSICAL LIMITATIONS: THE IMPACT OF HEPATITIS

In the meantime, I had become aware that hepatitis was a fairly common ailment in Southeast Asia during the early days of IVS activities there. However, the use of gamma globulin injections to boost immunity against the disease was not widespread. In addition, there was little awareness of how the disease was transmitted and why some persons were more susceptible than others. Early in my second IVS tour in Vietnam (1963-65), I was diagnosed with hepatitis. The white in my eyes turned yellow, my urine was dark, I had chronic diarrhea, and perhaps most debilitating of all, I had no enduring energy. The treatment then was complete hospital rest, non-fat diet rich in proteins and vegetables, lots of liquid, and no tobacco or alcohol, both of which had, up to that time, been an 'important' part of my life. After two weeks in the hospital, the symptoms had almost entirely disappeared, and I had renewed energy—as long as I didn't push myself.

Full of youthful over-confidence and heedless of medical advice, I soon reverted to my old lifestyle. I resumed smoking and drinking beer and my demanding work schedule left little time for sleep. I had no understanding of the need to control the pace of my activities and to listen to my body. As almost anyone could predict, I had a relapse. All my symptoms returned and I was back in the hospital for more rest. I hadn't been willing to realize how attentive a person must be to allowing his body, basically his liver, to repair itself. Since the liver is the only organ that can regenerate itself and each of us carries far more liver capacity than we normally need, the process was somewhat forgiving—up to a point!

In early 1965, I was wrapping upon my IVS team leadership responsibilities and transitioning to a contract with USOM/Rural Affairs as project manager of the "Summer Youth Program". Like the College of Agriculture's " Rural Youth Program", this project introduced urban-dwelling students from the University of Saigon to the rural countryside through local activities in villages around the country, not unlike our own living and working experience as IVS volunteers. The purposes of both of these programs were to familiarize students with rural life and to provide the opportunity to develop leadership in the process. My responsibilities entailed frequent travel, a lot of socializing, and participation in many activities, conditions that were right to earn me a second relapse, and I did, in spades! My doctors and nurses gave me stern warnings about the importance of rest, proper diet, and above all else, no beer or tobacco. This time the condition was extremely debilitating. On top of this, I was moving into a new position that required my full attention. I avoided hospital rest this time and decided I could treat it myself with proper planning: no outside activities, complete focus on the new job, getting ten to twelve hours of sleep each night. I kept up this routine for the duration of the six-month contract.

Later that year I returned to the States and spent the winter in Washington DC preparing for the new assignment in Thailand. Again, I found myself without strength, walking one hundred meters was exhausting. Rest and proper food were necessary. I would start to feel better but I found I had no endurance at all. I continued to prepare for Thailand and trying to regain physical fitness. When I reached Thailand in 1966, I was still far short of normal. I was determined to heal and I did everything I could to do so. I quit smoking, forever, and stayed away from alcohol, and rested prodigiously along with following proper eating habits. Little by little my endurance returned. I spent six years in Thailand and during that time, I believe I regained complete health.

As a result of this experience, I have no doubt, that one can change his habits, if he so chooses. The motivation to live a long and normal life was sufficient for me to stay the course to full recovery. I am not proud of my initial resistance to medical advice but over time, the illness and its treatment encouraged me to learn more about body health, nutrition and alternative healing procedures. For example, I learned about the use of milk-thistle as an herbal supplement that enhances the regeneration of liver tissue. I have taken it for at least fifteen years now, and my firsthand evidence convinces me it was the right and the important thing to do. This new interest in herbal medicine led to the collection of a vast amount of knowledge on health, nutrition and medicinal plants. This collection effort later included the development of a very rewarding relationship with the National College of Natural Medicine in Portland, Oregon. Ultimately, we donated many volumes from our large library to the College. The chain of satisfying experiences seems to never stop, but in the end, I have an original and unpleasant hepatitis infection from Vietnam days to thank for it all!

THAILAND: WORKING WITH THE BORDER PATROL POLICE (BPP)

In 1966, my career took a detour away from agriculture into the broader field of rural development, particularly as it related to rural security. At the time, U.S. involvement in the Indochina war was just beginning to intensify. The domino theory was in vogue. That theory held that if the three Indochina countries (South Vietnam, Cambodia and Laos) fell to the communists, the remaining countries of Southeast Asia would topple like dominos. Thailand was seen as potentially the next domino. U.S. foreign policy was dedicated to the preservation of the three Indochina countries as non-communist, but also to keeping the remaining dominos upright.

Recognizing the linkage between rural prosperity and the loyalty of the rural population to the Royal Thai Government, the U.S. Agency for International Development (USAID) was aiding the Government to promote rural security through rural development projects. This strategy later became known as the "hearts and minds" approach to remote populations. I was offered an interesting opportunity to join a new team being put together by the USAID in Thailand that would work on remote area security development. I believed that my Vietnam experience was highly relevant to this challenge and that Thailand was worth our best efforts to support its defense. I took the job. The program was aligned with the guidelines that evolved out of the Geneva Accords of 1954, the Indochina peace agreement following the French defeat at Dien Bien Phu. Under that agreement, Thailand would not place military personnel on its eastern borders. The Thai Government would set up a para-military police force for border security based in its Ministry of Interior. The Border Patrol Police (BPP) was the force created for this purpose.

During 1966-71, I had the distinct honor to work with outstanding BPP officers. Among the best were Major Wipas Wipulakorn, deputy commander for Region IX in the southern areas of the country including the dangerous border areas with Malaysia and Lt. Col. Somkhun Harikul, commander of Region IV headquartered at Udorn in Northeast Thailand. Lt. Colonel Somkhun was responsible for parts of the Mekong River border area with Laos. Both officers later assumed national leadership roles in BPP. Over time, Major Wipas was promoted to the rank of full general. He ultimately became commandant of the BPP and subsequently, commanding general of the Thai National Provincial Police for the entire country. Lt. Colonel Somkhun also rose to the rank of general and became Wipas's deputy at BPP headquarters.

Along with other American team members, I had months of intensive training in preparation for serving as a regional advisor in remote area security development with the BPP. Eventually, our team had one or more advisors within each of the nine regional police areas of the country. Home-grown and foreign communist agitators—not yet considered terrorists—were active in many areas of rural Thailand. Indeed, when I was posted in Songkhla in southern Thailand, I discovered that there were numerous threats to rural security. There were camps of the Communist Party of Malaysia in the jungles on the Thai side of the Thai-Malaysia border, agents of the Communist Party of Thailand infiltrating down the peninsula from the north, an Islamist separatist movement operating in Muslim areas in the south, and there existed a number of dangerous criminal gangs.

Within the BPP, counter-insurgency was already a reasonably well-understood function and was well underway in some areas of the country at the time of the USAID team's arrival. Over time, we developed strong working relationships with our senior Thai counterparts. Our task was to foster within the local border patrol greater emphasis on development activities for remote villages. With modest funding for development activities, we designed projects with village leaders to meet local needs and interests. Generally, we found ourselves working in one or more of three areas: education, agriculture and health. The goal was to inspire loyalty to the central government among these remote village populations.

Education projects were top priority as they had the backing of the Thai King and other members of the Royal family. These projects were initially located in villages that had been subject to some kind of political agitation. No money was provided directly to the project villages, but in-kind support was made available, primarily rice and related food items for villagers who agreed to work on school organization and construction. Wherever possible, local building materials were used such as rocks and timber. When a project was completed, members of the Royal family would arrive for its dedication, which greatly enhanced the project's status within the village. A BPP staff member would serve as the initial teacher for a period of not less than two years. Eventually, a local person would be trained to serve as the teacher. Subsequently, the school would be handed over to the Ministry of Education.

The Border Patrol continued school development projects under royal patronage for many years. While we thought the emphasis on education was a highly successful strategy for winning the hearts and minds of the villagers, we considered agricultural and family health activities to be about equal in importance.

We let the villages define their own agricultural projects. Most were interested in village gardens. BPP provided seeds and the villagers provided the labor. We also supported small irrigation projects, village tree planting and occasionally, the purchase of a piece of simple tillage or harvest equipment.

Village health projects included first aid stations, basic training for village medical workers, midwifery assistance, covering wells to reduce the risk of pollution of drinking water, and elimination of standing water to control malaria, which was a serious problem among remote populations.

Occasionally, we did village road projects that typically focused on improving drainage, forming the shoulders of the roadways, raising the roadbed, and building small bridges.

Whatever the project, each was intended to create positive relationships between the border patrol officers and villagers. We made sure that villagers always were invested in each project as manifested by their willingness to contribute at least their labor. Our development projects produced an essential by-product as a result of the cordial relationships the projects engendered: villagers became vital sources of useful information about the intentions and activities of local insurgents.

Responsibilities Of Marriage

One day in my second year with IVS in Vietnam, I visited the Asia Foundation to solicit funds for the tribal center program where I met, by chance, a very lovely and personable Vietnamese staff member named Simone. After a short conversation, I agreed to meet at an upcoming dinner at her family's apartment in downtown Saigon. We kept in frequent contact after that, but my job responsibilities required that I spend much of my time traveling up country in support of the agricultural team.

As time passed, I began to realize that I was attracted to her patience with my varying and unpredictable schedules and her gratitude for small favors. She was willing to accept any change of plans as quite normal, was always positive and was always in good humor. These qualities didn't escape my notice! During my second tour in Vietnam, which began in early 1963, I concluded that I must pay better attention to these sterling personal characteristics. We began to see each other a bit more and our relationship became more focused. But I still wasn't ready to settle down. Life in Saigon was most exciting and moving much too fast during those days in the early 1960s. The days and months flew by.

After I moved to Thailand to work with the BPP, I had time to think about our relationship. The separation from Simone was probably good for us as we began to understand what life was like without the daily presence of each other. I missed her and her delightful son. I stopped taking our relationship for granted. I felt a void in my life without their presence. By early 1969, I realized I was ready to accept the responsibilities of marriage. We agreed that we would get married in Saigon on 31 May 1969, an auspicious date determined by Vietnamese fortunetellers.

Following the wedding at the Catholic church on Tu Duong Street in Saigon and a grand reception and dinner at the Club Hippique with large numbers of family and friends, Phan thi Minh Chau Simone Anne Marie, son Paul and I travelled to our first home in Songkhla in southern Thailand where I was working with the BPP. This move was quite a change for both Simone and Paul. Simone had grown up in a well-placed Catholic family from North Vietnam and had emigrated to Saigon in 1954. Her father had been a well-connected lawyer who served the French colonial government. Her older brother, Jacques, a Saigon businessman, maintained the family residence in Saigon where she originally lived after their father passed on soon after moving south in 1954. Two other brothers, Rene and Louis, were living in France.

Suddenly, married life meant new responsibilities for me. At age thirty-one, it was probably about the right time in my life—before I became too set in the ways of living as a bachelor. I saw myself as very lucky to have found such an engaging wife.

Simone home-schooled Paul during our first year in southern Thailand, no easy task as it was all done via the Calvert system in English. Paul made rapid progress in his studies including learning the English language. Also, at the age of seven he was playing with Thai children and learning to speak

Thai in short order. For the second year we felt Paul needed the opportunity for social development that could be provided by a school. With the home stay assistance of fellow USAID teammate Bob Plummer and his wife Sue, Paul spent his next two years attending the international school in Chiang Mai, northern Thailand. He did well in academic work and Bob and Sue were excellent foster parents.

BACK TO VIETNAM

After I finished two tours in Thailand with the BPP, we returned to Vietnam in 1972. This time, we were on our own. I had no idea how events would play out over the next three years. I was hoping to become a permanent ex-patriot and I wanted to get back into agriculture. However, for the next three years employment opportunities related to the war continued to divert me from my primary interest.

Simone's family had been encouraging me for some time to establish a company to bid on military contracts. Our company successfully completed two contracts for boat repair and truck transportation. However, the work required a degree of deviousness to satisfy U.S. government bureaucrats, and I soon realized that this kind of venture wasn't for me. By mutual agreement within the family, we closed shop. I did take away from these ventures important insights into the contracting process, the complexity of contract documents, and the games that are played by the parties in the course of contract implementation. This experience would serve me well when I was contracting with growers in Oregon to produce vegetable seeds a few years later.

In 1973, I moved on to a job with Computer Sciences Corporation (CSC). The firm had a contract with the Office of Field Operations of the U.S. Embassy in Saigon to operate a program called the Hamlet Evaluation System (HES). HES was designed to provide data on the security and economic situations in every village in South Vietnam. My previous IVS experience in the central highlands was the main reason I was hired for this job. Consequently, I was assigned to cover II Corps (read "2 corps"), the second military region, which included all of the central highland provinces.

Vietnamese villagers and provincial administrators provided the raw data for the HES system on a regular basis. I was a member of a team of four regional analysts who were expected to interpret the mass of information generated by the HES system to determine why any changes had occurred at the village level and report accordingly to the U.S. Embassy. We had to process and review all the data before it was entered manually on punch cards into an early computer system. HES work gave me considerable insight into the deteriorating security situation in II Corps. As the situation worsened, I began to realize that I was losing control over my future.

THE MILITARY DEFEAT OF SOUTH VIETNAM

Back in January 1973, U.S. Secretary of State Henry Kissinger and his North Vietnamese counterpart, Le Duc Tho, had signed the peace treaty that theoretically ended the Vietnam war. The treaty was in reality a fig leaf covering the withdrawal of American combat forces. (Kissinger and Tho were subsequently awarded Nobel Peace Prizes.) As part of the U.S. military drawdown that occurred over the next year, the HES program—including my job—was terminated.

The peace accords ended the direct role of the U.S. military in Vietnam. However, the Defense Attaches' Office (DAO) of the U.S. Embassy in Saigon continued to monitor the security situation. The DAO appreciated my HES background and, in mid-1973, offered me a contract position as liaison officer with the South Vietnamese army headquarters back in II Corps at Pleiku. There I worked with two other persons assigned to the region, John Good and Nelson Keefe. Our task was to obtain reports on security conditions and military actions from the Vietnamese army (the Army of the Republic of Vietnam or ARVN) and pass them back to the Defense Attache at the embassy. In other words, we were trying to obtain the same kind of summary information that the now-departed U.S. military had been collecting except that we were obtaining information solely from the ARVN.

Despite the peace treaty, the North Vietnamese Army (NVA) rebuilt its forces and continued to fight. By early March of 1975, the NVA was steadily advancing southwards and the military situation in II Corps was rapidly worsening. On the night of March 14th, the NVA bombarded the Pleiku military headquarters with artillery fire and rockets. The fuel dump was set ablaze illuminating the night sky.

Early on the morning of March 15th, when I arrived for work, the II Corps headquarters was deserted except for a couple of dedicated senior Vietnamese staff officers. The II Corps Commander and the remainder of his staff had departed in the night to Nha Trang, a stronghold on the coast. Their appreciation of the security situation was crystal clear: the NVA were on their way. We needed to get out too, but our work wasn't done. There was a large number of vulnerable Vietnamese—employees of American organizations and their families—who could easily be subject to reprisals when the NVA took over. Earl Thieme, who was the provincial representative of USAID and the only other American official left besides the three of us, took charge of organizing the evacuation. Though the CIA team had also left on the first flight in the morning, Air America, the CIA's contract airline, which served all U.S. government organizations, continued to service our location. Earl called for transport and soon silver World War II vintage C-47's, the same type of aircraft that flew paratroopers to drop zones over Normandy on D-Day, landed at our airstrip. Nelson and John rousted the Vietnamese employees and their families while I became loadmaster at the airfield.

Pandemonium reigned at the airstrip as the frightened Vietnamese scrambled to get on the first plane. My shirt was literally torn off my back by the swarm of panicked evacuees. With a loud voice and a defiant glare, I tried to restore order. The crowd took my broken Vietnamese seriously, much to my surprise, and accepted a system of orderly departure.

The Air America C-47 was configured to carry up to thirty-two passengers. That day we loaded three C-47 flights and one larger C-46. I estimated that there were about 250 people on the four flights—more than double their normal capacity. Late in the evening, I got on the last flight accompanied by a German Shepard that had been our mascot. All the flights flew to Nha Trang which had become the new II Corps headquarters.

The DAO re-assigned me to Tuy Hoa, the capital of Phu Yen, the province immediately to the north of Nha Trang. Chuck Brady, the USAID representative, and I stayed for the next ten days in the now-empty CIA compound where we had access to excellent communications gear. I continued to liaise with the ARVN. Chuck and I tried to keep Saigon informed about the security situation that was becoming more dismal every day. On the night of March 30th, the NVA launched a rocket attack so devastating that we called for help. We were told that we couldn't be extracted until morning. We survived an otherwise quiet night, and an Air America chopper duly arrived at daybreak and took the two of us back to Nha Trang.

The official U.S. position, or as it turned out, official wishful thinking, was that the coastal enclave at Nha Trang would hold. However, the withdrawal of ARVN troops from the central highlands had become a general retreat with civilians and military personnel jamming the roads heading south. At the end of March with the fall of Da Nang, an estimated 100,000 South Vietnamese soldiers surrendered after being abandoned by their officers.

At Nha Trang on April 4th, I was still working at the U.S. Consul General's office when I looked out the window and saw a mass of ARVN soldiers pushing their way into our local PX which they were about to loot. It was again time to leave. Soon a helicopter landed in the compound to fly the consulate personnel and me to the airfield. The South Vietnamese Air Force and Air America were evacuating all and sundry to Saigon. We tried to make sure that our Vietnamese employees were getting out, but I was a newcomer to Nha Trang and didn't know the people well so I couldn't provide much help. I remember a scene from the chaos when a four-engine Air America C-54 revved its engines to get separation from the crowd and blew people and their bags back perhaps thirty yards. I boarded the first C-47 with space available and was in Saigon in a little over an hour.

After the debacle on the coast, the official thinking was that the situation in the South would finally stabilize. However, soon after I got to Saigon, I made sure that my wife, Simone, and son, Paul, and our most valued possessions got out immediately. They left in early April for Bangkok. All of our personal possessions made it out as well except for two antique Citroen roadsters, which I had been restoring in my spare time.

Despite the chaos of these weeks, I still had a job with the Defense Attaché's Office. I was assigned to work on evacuation issues at Tan Son Nhut airbase, the former U.S. military headquarters for South Vietnam and the main airport for Saigon. The Embassy wanted to make sure that the Vietnamese employees of U.S. organizations and their families would be taken out of the country to safe havens if at all possible.

I worked at Tan Son Nhut from April 5th to the 23th by which time the writing on the wall had become perfectly clear. I decided that it was time for me to leave Vietnam and flew to Thailand on the 24th. I found my wife and son in the apartment she had rented in Bangkok. Vietnamese friends and members of my extended family besieged our apartment, eagerly seeking news of the situation in Saigon. They pleaded with me to do what I could to get their brothers and sisters, sons and daughters, parents, grandparents, uncles, aunts—anyone they were related to—out of the country as they feared for their lives. They bought me a return ticket to Saigon on what turned out to be the last Air Vietnam flight into the South Vietnamese capital. On April 28th I was back in Saigon. While the situation was verging on chaos, it was hard to imagine that the iconic photos of the rooftop evacuation from the American Embassy would be taken in less than forty-eight hours.

Harlan Grosz, an old friend who also had a Vietnamese wife, had returned to Saigon with me on the same flight. He was on a similar mission to find and bring out family members. When we got in touch with each other, we had both concluded that our efforts were futile. Saigon was in chaos and it was impossible to find or contact family members and friends. The North Vietnamese were moving closer, bombs were being dropped on the riverfront from South Vietnamese aircraft seized by the North, and a number of structure fires were raging out of control. The orderly evacuation process had completely broken down.

On the morning of the 29th, Harlan called me from the airport and told me that the U.S. Military Assistance Command advisors were leaving and said he thought we should get out as well. I found a U.S. Mission car with keys in the ignition and fuel in the tank and drove to a designated pick up point on Cong Ly Boulevard. A military bus full of Americans soon arrived and I was given a flack jacket and a sawed-off shotgun for the ride to Ton San Nhut. Vietnamese soldiers were still guarding the gates to the airfield and they tried to stop our unmarked bus but our driver panicked and hit the gas. We sped through the gate as the soldiers fired into the air. As we got off the bus, we were relieved of the shotguns and herded into the DAO library to wait for a flight out. By chance, I picked up a copy of Edwin O. Reischauer's *Beyond Vietnam*, an authoritative critique of the U.S. adventure in Vietnam, to read as we waited. Indeed, I was soon to be "beyond Vietnam", physically, psychologically and vocationally.

Three "Jolly Green Giant" helicopters landed, each carrying fifty marines, to secure the evacuation area. These choppers were our ride out and I got the last seat on one of them. As we spiraled straight up to avoid ground fire, I could see fires burning throughout Saigon and suburban Cholon. After flying for about twenty minutes, we crossed the coastline. A cheer went up from the passengers. Over water, we were out of reach of any hostile ground fire. We landed on the USS Vancouver, the amphibious transport ship that had dispatched the marines to Saigon. The ship lingered off the coast of Vietnam and I read Reischauer to pass the time. Soon, the first of the boat people began to arrive. These people had fled in panic with only the clothes on their backs and they were hoping that the U.S. Navy would pick them up. Some of the larger ships did so but the Vancouver was already overcrowded with evacuees from Tan Son Nhut. Her sailors provided food and water to the boaters and saw them off on their way. I was on the Vancouver for a week before the ship docked at Subic Bay in the Philippines. I called my wife in Bangkok who was understandably concerned about what had happened to me. I provided reassurance that I had 'kept my ears low' during the final days of Saigon's independence.

AFTERMATH: BEGINNING AGAIN

From Subic Bay, I travelled by coach to Manila, where the American Mission was busy meeting the needs of the many new arrivals from Vietnam. I was moving around in a bit of a daze, trying to understand what had just transpired in the last few weeks. I spent some time wandering around Manila and met a number of old friends including Bert Fraleigh of USAID, who was able to arrange lodging for me at the Army-Navy Club. I looked up Jim Rinella with the Dole Company to check on the possibility of a job. However, I wasn't sure what I was looking for and was just exploring leads. All the ex-pats from Vietnam seemed to be doing the same thing. I decided that I might contemplate the future in Bangkok as well as in Manila. So after a few days, I flew back to Thailand to be with my wife and son.

In the previous six weeks, I had been evacuated four times ahead of the advancing North Vietnamese army. The fall of Saigon not only ended the war, it closed off my option for working in Vietnam. As I reflected on my Vietnam experience and thought about the future, I also realized that a phase in my career was over. I knew it was time to move on, but I had no idea what direction I wanted to take. It was a very uncertain time for me, as any career change might be for anyone.

> *The important thing is that when you come to understand something, you act on it, no matter how small that act is. Eventually it will take you where you need to go.*
>
> Sister Helen Prejean

Simone, our son, and I started over by moving to Oregon close to family members in the Pacific Northwest. With patience and attention to what I understood and where I found myself, I was able to begin again—literally, a beginner's posting with Agricultural Services Corp.—in which I gradually found my place in agribusiness with Agricultural Alternatives, Ltd. and a very rewarding 'second' life. It came to me—as I have described in the preceding pages—one step at a time.

Top photo: Squatting center front left, Mr. Chuong Hsiung Yu, JCRR horticulturist. Squatting center front right, Mike Chilton, and standing center, Don Wadley, both IVS members assigned to the Vegetable Research Station, Dalat, 1961.

Bottom photo: Most of the new crop development work on white and yellow potatoes took place at Dalat. Climate conditions were very favorable for this crop.

Top photo: Mr. Binh (farmer) and Mr. Dinh (extension agent) with first overflowing bulbing onion harvest drying in Mr. Binh's home, Dalat, Vietnam, 1961.

Bottom photo: Tremendous personal strength is apparent in rural children in areas where water pumps were still relatively scarce. Here, a young girl carries two 5-gallon containers of water, or approx. 80 lbs.—near or exceeding her own weight. Nguyen van Quy, interpreter, and Don Schmidt, Dalat IVS member, are in background.

Vietnamese technician inspecting bulbing onion trial at Horticultural Experimental Station in Dalat, Vietnam, circa 1961.

Mike Chilton (IVS), Dr. Paul Ma (head, JCRR team from Taiwan) and Joe Hamilton (USAID horticultural Adviser) at Dalat Horticultural Experiment Station, Vietnam, circa 1961, confer on the production results of some of the first Irish potatoes grown in the South Vietnam highlands.

Don Schmidt, IVS member in Dalat, was always interested in new crop adaptability. Here, Don examines a rather successful trial planting of fiber flax. Unfortunately, fiber flax was a crop without an immediate market.

Mike Chilton prepares a seedbed to produce seedlings for a variety production trial of bulbing onions at the Dalat Horticultural Research Station, 1961.

Bulbing onion field trials near Dalat. Left to right, Mr. Dinh, provincial extension, Khiem, interpreter, Mr. Chuong Hsiung Yu, JCRR technician, Kinh, interpreter, and Don Wadley, IVS, 1962.

Top photo: Dalat Horticultural Reasearch Station Headquarters, Dalat, 1960.
Bottom photo: Tribal Village near BaoLoc, 1962.

An early commercial production of Irish potatoes near Dalat. Vietnamese workers dig the harvest, and Bao Loc IVS member, Bob Knoernschild looks on from the right.

Top photo: Tribesmen carrying to market cinnamon bark harvested wild in the mountains west of Quang Ngai city, Central Vietnam, circa 1962. At the time, spice markets
around the world considered cinnamon originating from the hills of Vietnam to be
some of the best quality anywhere.

Bottom photo: Tribal women and children in Lam Dong Province, circa 1962.

IVS/Vietnam team leadership,---Don Luce, Chief of Party/Vietnam, Jim Kelly, Deputy COP for Admin., and Mike Chilton Highland Ag. Team Leader—with Filipino nurses who worked at the SDA Co Doc Hospital in Saigon, and where IVS members received their medical care, circa 1962.

On arrival at the Dalat assignment (August 1960), I met the following resident staff at the IVS house, left to right: Kham, driver; Quy, Interpreter; Ong Bep, our cook; and Chi Hai, the house maid, all provided my first exposure to the issues of "personnel management" —-later to be realized as an unexpected bonus of the IVS experience!

IVS Headquarters, Ton Son Nhut, Saigon, 1961.

David Nuttle, IVS member working with the tribal people in Darlac Province, demonstrates field working attire of the Rhade people from the Ban Me Thuot area, 1961.

IVS gathering with Vietnamese staff and friends in front of IVS headquarters, Ton Son Nhut, Saigon, circa 1962.

In the early days of IVS/Vietnam, IVS team members enjoyed a productive and amiable working relationship with US military advisors. Here, we see the entrance to the MAAG compound of Ban Me Thuot. The structure, lavishly built as three over-sized Montagnard longhouses, had formerly been the hunting lodge of Emperor Bao Dai during his visits to the Highlands.

Iowa City mom relieved; son now out of Saigon

By M. JOANNE BRUEGGER
Of the Press-Citizen

When President Ford ordered the evacuation last Tuesday of all Americans from Saigon, one Iowa City mother thought she had cause for relief.

Verna Chilton learned Sunday, however, that her son, Michael, had not been among the first-of-the-last evacuees, as she had thought.

Michael Chilton, a member of the U.S. defense attache's staff in Saigon, did not leave the country until Thursday. He telephoned his mother from Manila Sunday morning.

Mrs. Chilton, of 22½ South Dubuque Street, had not heard from her son for several weeks before his evacuation. She

did know that his North Vietnamese wife, who had fled the Communist regime, and her 13-year-old son had been removed to Bangkok, Thailand, in March. Chilton told his mother Sunday he would be joining his family in Bangkok, but currently "things are very uncertain."

Mrs. Chilton said her son first went to Southeast Asia in 1960, with the International Voluntary Service. He has remained in that area since, and last year took a position with the defense attache's office.

She said Chilton told her only that he had come to Manila by boat, that it had taken several days, and been "a rough trip."

If he encountered difficulty in leaving the city, he did not tell his mother. "He wouldn't have dared," she said.

MICHAEL CHILTON

Local newspaper clipping, Sigourney, Iowa, following the downfall of Saigon, Vietnam, April 1975.

SEVEN

CONTINUING INTERESTS

*W*hile *closing Agricultural Alternatives* and finalizing transfer to the new owner, I was exploring options for a useful occupation following my adventures in the seed world. Obviously, one targets one's own interests, but often details develop quite unexpectedly and sometimes serendipitously.

MY LIBRARY

I began collecting historical and strategic books during my early years abroad, particularly as they related to my work. Subject matter varied: Far Eastern history, insurgency, affects of conflict and reasons for inevitable change, and occasional books on wildlife and cultural aspects of the countries and regions where I was working.

But not until we returned stateside in 1975 did the groundwork for the current collection begin to take shape. After I had decided that my best career direction was to pursue some aspect of agriculture, it occurred to me that specialty production of medicinal plants would be exciting, rather unique, and much in keeping with my ongoing interests in natural science. I had not yet realized the impracticability of growing medicinal plants without a good knowledge and capability of processing and markets, not to mention my total unfamiliarity with the entire spectrum of production.

Nonetheless, I continued to nurse the dream during my first years of re-familiarization with U.S. agriculture while I was working for Agricultural Services Corporation. My searches for medicinal plant literature on library shelves at institutions of higher education in the Pacific Northwest generally met with disappointment. In the meantime, at Ag Services, we were beginning to grow limited amounts of vegetable seed for international customers. This work led to one or more major international trips each year to negotiate with customers about future orders. These trips also provided excellent book exploring and accumulation opportunities.

It is important to note here that the vegetable seed and other specialty crops we grew were all grown under proprietary contracts, i.e., grown for a specific market from parent seed derived from a breeding program. Medicinal plants, on the other hand, were grown as commodity crops and were often available from several sources. Crops grown under proprietary contracts provided big marketing advantages for a small company. In spite of this advantage, I was beginning to realize that medicinal crops were potentially a golden opportunity. In the course of my travels, I re-doubled my efforts to explore the literature on medicinals from various parts of the world. Indeed, I soon learned to find more material on medicinal plants abroad than in the States.

I began buying and dragging home suitcases full of books, occasionally finding rare and old gems in the process. My book collection became a serious hobby and provided another reason to explore wherever I went. Though focused on agriculture, my areas of collection interest were expanding to include: early agriculture, its history, its implements, its crops and buildings; botanical material of all kinds: herbals, botanical history and its explorers, new crops and marketing systems, economic botany, history, and references on the avian world of poultry, waterfowl, upland game birds. (An interest in avian studies was a holdover from raising and showing many kinds of poultry and birds as a child growing up in Iowa).

My combined travel and collection continued for well over twenty-five years. As I collected, I also learned about antiquarian book fairs, particularly in the United States, which are very exciting places to learn what was on the market and to make new friends. As my collection grew, I developed a network of friends that shared common interests. These relationships were sometimes costly, sometimes yielded great bargains, but overall, were invigorating and perpetuated the continuation of a hunt that would never seem to end. The collection expanded to over 8,500 volumes throughout our residence and was eventually computer cataloged. During earlier times, it never occurred to me that someday something would have to be done with the collection either under my direction or within my estate! And yes, I definitely prefer the former!

THE NATIONAL COLLEGE OF NATURAL MEDICINE (NCNM)

One day while making database entries of accessions into my library, a Portland State library sciences student, Alicia Yokoyama, suggested I might find interesting the collections of the NCNM library. I followed her advice and visited the NCNM facilities in Portland. There I met Susan Hunter, NCNM Vice President who arranged several opportunities to learn more about the school, staff and activities, and to participate in a number of social activities.

My wife and I enjoyed very much this interaction. Subsequently, I was invited to join the NCNM Board of Regents for the purpose of bringing my outside interests in natural health to the school. I felt very honored and have continued to sit on the Board for the last five years. Our relationship with NCNM has grown into a very satisfying and pleasant experience. Of more immediate consequence, I realized that the school was the proper place for my extensive library collection on medical botany. The NCNM library offers the best opportunity that I know to house and expand access and utilization to our collection. We worked from 2010 to 2014 with NCNM librarian, Dr. Rick Severson, transferring relevant volumes to the NCNM library. Our contribution to the library comprises over two thousand volumes. Valuable antiquarian titles are safeguarded on locked shelves in a specially created reading area within the library. We believe and we hope that this arrangement will prove to be a better way of utilizing years of pleasant collection efforts offering continuity and growth, far outliving our own personal interests and capabilities.

An article on the library donation is re-printed from the NCNM newsletter as **Appendix Three.**

OREGON COMMUNITY FOUNDATION (PORTLAND, OREGON)

Our lawyer, George Jennings, introduced us to the Oregon Community Foundation (OCF), a forty-year-old, state-wide community service fund, as a way to help plan for retirement and put aside resources for ideas and projects that we might wish to undertake in retirement. We ultimately set up the "Mike and Simone Chilton Family Advised Fund", serving as a useful receptacle for certain estate resources.

During the establishment of the Fund, OCF asked whether I might like to serve as a volunteer grant evaluator. After learning more about the role of a volunteer evaluator and sitting in on related meetings and training sessions, I decided to volunteer for the semi-annual evaluation sessions. Now completing my seventh year as evaluator, this activity has offered eye-opening insight into local issues of social and economic consequence that I had neither not known about nor understood, as well as opportunities to meet interesting people.

ANTIQUE POWERLAND MUSEUM ASSOCIATION (BROOKS, OREGON)

Antique Powerland is a collection of fifteen educational museums located on a sixty-two-acre campus at Brooks, just north of Salem. Each museum focuses on aspects of early technology and power as they have contributed to development of the Pacific Northwest region over the last 150 years. The museum complex has been at its current location for over forty years. It is entirely supported by private resources and has been totally operated by volunteers.

While I had known of Antique Powerland for over thirty-five years, my direct involvement began six years ago when I became a member of the board of directors. Now a former board member, I have for the last six years held assorted committee positions and have worked on various projects to upgrade museum facilities, improve landscaping and enhance visitor-friendliness.

International Voluntary Services (IVS) Alumni Association

I joined IVS over fifty-five years ago immediately out of college. IVS took me abroad to begin my overseas career. I have maintained a commitment and familiarity with the organization and what it stands for. Many hundreds of former volunteers maintain the friendships of that early period of their lives. This early volunteer experience in the international development field provided an operational format that was copied and used extensively during the formative days of the Peace Corps. The IVS Alumni Association (IVSAA) acts as an institutional memory for former volunteers and facilitates interaction of those who wish to remain informed about associated activities. I have served as president of the association since 2003.

Since 2010 a group of alums have been interested in documenting the IVS experience and contributions over the past half century. Their effort resulted in the 2015 book publication of *The Fortunate Few*, *IVS Volunteers From Asia To The Andes*, by Thierry J. Sagnier, published by NCNM Press.

SeedAsia, (Bangkok, Thailand)

SeedAsia is a seed company founded by a long-time friend, Tim Welsh. The firm is headquartered in Bangkok, Thailand. The company has been operating for approximately ten years and has developed its own hybrid seed corn varieties for local adaptation, both domestically and regionally. We were invited in 2010 to join a group of local and expatriate investors to purchase shares in the firm. With our minor investment, we were able to participate in the rough and tumble growth of a relatively new crop in that part of the world. After an initially slow start, the company become profitable and began making long-delayed capital investments. We monitored with interest how this very young company addressed the trials and tribulations of decision-making and how it fared in that opportunistically challenging developmental environment.

More than a year ago, the company was purchased by a major European seed company. At that time all shares were redeemed from the original shareholders, meeting the requirements and appreciation of all concerned.

APPENDIX ONE: THE WILLAMETTE VALLEY

Willamette Valley - the Perfect Growing Environment

When the Willamette Valley was created, the perfect environment for seed production was made. Located 30 miles inland from the Pacific Ocean between the Coast Range on the west and the Cascade mountains on the east, the Willamette Valley enjoys fertile soils and a year-round temperate climate.

The Japanese current brings warm waters along the Pacific Northwest coast which helps give the Willamette Valley a moderate winter and a dry summer. The Valley receives an average annual rainfall of 1000 mm. This rain falls mostly in the winter months, with July and August receiving very little precipitation. Moderate winters allow an excellent environment for growing biennial crops which are planted in the fall and carried through the winter for vernalization. Dry summer months offer excellent harvesting conditions.

With little precipitation and low humidity during harvest time, the crops are dried naturally in the field to proper storage moisture content. No additional artificial after-harvest drying is usually necessary. This means excellent quality seed, both in viability and appearance.

Quality seed production in Oregon

A Long Production History

Seed production is an important industry. Willamette Valley growers have many years of experience and understand the production of quality seed for world markets.

Quality seed production in the Willamette Valley began at the turn of the century and increased rapidly as the world's transportation network improved. Vegetable and specialty seed production dates back to the 1930's. Since then, all temperate season crops have been grown including: Brassicas, Radish, Squash, Pumpkins and many others.

CURRENT SEED CROPS

Cabbage	Long Day Onion
Brussels Sprouts	Bunching Onion
Turnips	Garlic & Leek
Rutabagas	Summer Squash
Parsley	Winter Squash
Parsnip	Pumpkins
Sugar Beets	Radish
Red Beets	Chinese Cabbage
Swiss Chard	Oriental Brassicas
Medicinal Plants	Grass Seed
Spices	Legume Seed
Herbs	Mustard
Flowers Seed	Kale

Efficient Transportation Networks

Shipping can be handled efficiently and speedily regardless of the final destination. The Port of Portland, located at the northern end of the Willamette Valley, is one of the busiest shipping ports on the west coast. All major steamship lines are represented in Portland and offer services to ports-of-call around the world. Portland is the last port-of-call for vessels with destinations in Asia. Transit time between the Port of Portland and Tokyo is only ten

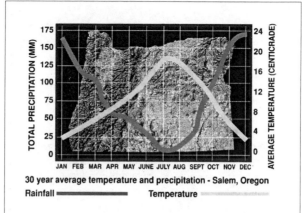

30 year average temperature and precipitation - Salem, Oregon
Rainfall ══════════════   Temperature ▨▨▨▨▨▨▨▨

days, the shortest of any west coast port.

Transit time for European ports is approximately 30 days by ship. If delivery time is critical, a land bridge service is available. A container can be shipped by rail to an east coast port where it is loaded onto a vessel, cutting the transit time by one-third.

Portland International Airport offers air transportation to all domestic and international destinations. Daily flights are available to several destinations in Asia via Seattle, only 35 minutes away by air. Several daily flights to Europe and beyond are also available.

With an excellent interstate highway system, the Willamette Valley is only a few hours away from all major cities on the west coast.

Research Assures Quality

Oregon State University (OSU), located in Corvallis, Oregon, in the center of the Willamette Valley offers degrees in many areas of agriculture including seed production. It also provides research to assist growers to improve production technology, seed yields and to produce quality seed crops.

Acting as a liaison between OSU researchers and growers is the Oregon State University Cooperative Extension Service with crop specialists in each county. The Extension Service provides technical advice to growers, and the latest production techniques through meetings, publications, and farm visits.

Another activity of the OSU Extension Service is seed certification. Field crop seed produced under the certification program is grown according to a uniform set of rules used throughout the nation for isolation of genetic and physical purity and viability. Seed certification also oversees production of seed under the OECD scheme, a worldwide program to ensure the production of quality seed.

Radish seed field being harvested in the Willamette Valley

OSU operates a nationally recognized Seed Laboratory which conducts purity and germination analysis for all certified and OECD seed lots produced in the state. It also conducts tests on thousands of vegetable seeds, specialty seeds, and field seeds not produced under the certification program.

Phyto-sanitary requirements for seed marketed internationally are met through the inspection process by Oregon Department of Agriculture guidelines. These inspections assure that seed quality meets the standards of international markets.

Specialty Seed Committee

I n 1980 several companies involved in vegetable and specialty seed production formed the Willamette Valley Vegetable and Specialty Seed Committee. This group with the assistance of the Oregon State University Cooperative Extension Service, was formed to promote quality seed production. One of the major activities is to maintain maps where fields are marked and recorded to ensure adequate isolation distances between crops and to ensure buyers seed that is true to type.

Other activities of the committee include herbicide trials in conjunction with the weed specialists from Oregon State University, and promotion of the Willamette Valley as a quality seed production area.

**Developed by Willamette Valley
Specialty Seed Crops Association
in cooperation with
Oregon State University Extension Service**

For further information contact:

**Oregon State University Extension Service
Department of Horticulture
4017 Ag & Life Science
Corvallis, OR 97331
Phone: (503) 737-5461
or (503) 737-3464**

APPENDIX TWO

AGRICULTURAL ALTERNATIVES PRODUCTION AND MARKETING AGREEMENTS

Production Agreement, 1996 Harvest

Agricultural
Alternatives
L I M I T E D

No. 61 Japanese Greens

Location: Willamette Valley

P.A. Contract No: 636

Date: April 12, 1996

Lot No: 6IT-05RO

(503)873-6267

Marketing Contract No: 663

This Document is a purchase confirmation of the product described below in the quantity, quality, and upon the terms and conditions as provided herein to:

Producer:

Ray Kuenzi
2834 Hibbard Rd., NE
Silverton, Oregon 97381

Acres:	Product:	Specifications :	Grower price:
Production of __5__ Acres	No. 61 Japanese Greens	−Purity: 99% −Germ: 85% −Moisture, 8% −Noxious weed free. −Grain free. −Meet All Japanese import standards.	$ _$0.80_ lb. Cleaned Wt. & bagged

Shipping Information:
FOB/Warehouse: Cleaning warehouse
Shipping date: When Crop meets above specifications or not later than December
31 of crop year.
Packing: 55 lb. bags or as requested. Bagging to be paid by buyer.

Terms of payment: 30 days from date of shipment.

Remarks: −To be planted as early as weather permits in the spring.
−Stockseed @ $10.00/acre payable at settlement.
−If the field is removed without prior approval, the stockseed is due
and payable.

Approval/Acceptance

Producer: Ray Kuenzi   Buyer: Agricultural Alternatives, Ltd.

By: _____   By: _____

Executed in duplicate, one copy to be signed and returned immediately.

3533 Ridgeway Dr. S.E. Turner, Oregon 97392 USA
Phone (503) 743-4338   Fax(503) 743-2451

Marketing Agreement, 1996 Harvest

Greens No. 61

Agricultural Alternatives LIMITED

M .A. Contract No : 663

Producer Contract No:_____

Date: March 13, 1996

This Document is a purchase confirmation of the product described below in the quantity, quality and upon the terms and conditions as provided herein to:

Purchaser: American Takii, Inc.
301 Natividad Road
Salinas, California 93906

Quantity:	Crop:	Specifications :	Price:
Production of : Greens (OP) 10 Acres No. 61 AC 5		−Purity: 99% −Germ: 85% −Moisture, 8% −Noxious weed free. −Meet All Japanese import standards.	$ 1.20 /lb. Cleaned Wt. & bagged :

Shipping Information :

FOB / Cleaning warehouse

Shipping date: When crop meets above specifications or not later than
December 31 of harvest year.

To: To be advised   Routing: To be advised

Packing: 55 lb. bags or as requested. Bag Costs to buyer's account.
Terms of Payment: Cash on receipt of documents.
Bank: To be advised

Remarks: −Stockseed to be provided without charge.
−FIS rules to apply.

Approval/Acceptance
Purchaser: American Takii, Inc.   Seller: Agricultural Alternatives, Ltd.

By: _____   By: _____

Date: 4/2/96

Executed in duplicate, one copy to be signed and returned immediately.
3533 Ridgeway Dr. S.E.   Turner, Oregon 97392 USA
Phone (503) 743-4338   Fax (503) 743-2451

Appendix Three
National College of Natural Medicine Library Donation

DONATION OF RARE BOTANICAL BOOKS ENRICHES LIBRARY

As you enter through swinging glass doors, the NCNM Library seems like a typical small college library: the hushed whispers breaking the respectful silence, the unforgiving glare of fluorescent lighting shining down on students in study carrels, heads bent over books, intent on classwork. But a short walk from the second set of glass doors brings you to a closed door next to the library stacks—and behind that door is another world entirely.

NCNM's Rare Book Room houses a collection of 1,500 rare and precious volumes on naturopathic and herbal medicine, worn and carefully thumbed through by generations of medical scholars who relish the opportunity to glean invaluable—and nearly lost —knowledge from practitioners who have long since departed.

The NCNM Library, already one of the largest and most unusual collections on natural medicine in the United States, is about to grow significantly, thanks to Michael and Simone Chilton. The couple is donating 1,500 books, almost all of them about botanical plants, many of them extremely rare and valuable.

And Rick Severson, PhD, NCNM's normally reserved library director, can't contain his excitement about the collection. "This is one of the most important gifts our library has ever received," he said, adding that a special area will be created to house some of the most valuable items in the collection, which includes books dating to the 17th century.

Mike Chilton, a member of the NCNM Board of Regents, has been growing his collection since 1975, beginning with his journeys throughout Southeast Asia. Chilton, who holds a master's degree in botany, was teaching agriculture overseas when he developed an interest in vegetable seeds and medicinal plants. He soon realized that, while university libraries could provide some books, his deepening curiosity required cultivation of his own collection of reference materials.

"My collection developed during my world travels," Chilton said of the years during which he often discovered written gems in shops and collections. "I didn't have a crystal-sharp idea of what I wanted. I took more of a shotgun approach—imperfect, but surprisingly effective in satisfying my desire to learn about medicinal plants."

Chilton found himself investing substantially and, eventually, noticing a business opportunity. "Medicinal plants were already a commodity," he said, "but vegetable seed hadn't yet grown into a commodity." He claimed the niche in 1985 by starting his own company, dedicated to growing and selling vegetable seeds.

> "These books contain information that is just so hard to come by. There are only about 30 libraries in the world, for instance, that have information about African poisonous plant, and NCNM Library is one of those."
>
> —Rick Severson, PhD

Severson said that when the Chilton collection is complete at NCNM, it will contain invaluable knowledge that had nearly been lost. Chilton agrees: "Botany was so important to pre-modern science; it pre-dated the Linnaean classification system of herbs and plants that developed with modern science. There are a half-dozen books in my collection, for example, like the 1759 edition of *The Gardener's Dictionary*, by Philip Miller, which are extremely valuable and typically can only be found in a botanical collection, like that of the Lloyd Library in Ohio."

"These books contain information that is just so hard to come by," Severson added. "There are only about 30 libraries in the world, for instance, that have information about African poisonous plants, and NCNM Library is one of those. It's an incredible synchronicity that this collection appears just as NCNM begins its Post-Graduate Certificate in Botanical Medicine program."

Chilton said he expects the gift to NCNM will draw people to the program. "NCNM seemed to be the best place to share our collection," he said. "I know that there's a better opportunity here for the books to be used for research and education."

Left to right: ND student Hunter Peterson, Dr. Schleich, Mike Chilton, Susan Hunter

NCNM Library participates in a national interlibrary loan system (which has attracted the attention of libraries internationally), and Severson expects many of the Chilton books will be of interest to the growing number of botanists in the region. "The rare books from the 1600's will be kept under lock and key," he noted, "but some of the other books will be available for interlibrary loans."

The exquisite calligraphy and hand-drawn illustrations within the books about folk medicine and plants are typical of books of that era. Some of the works in the Chilton collection, Severson said, are known as "incunabula": volumes printed in Europe before 1600, during the period he called a "gray zone" in the history of book printing, when production was in transition from handwritten manuscripts and block printing to movable type on the early printing presses.

No account of the NCNM Library collection is complete without a nod to Severson's predecessor, Friedhelm Kirchfeld, who shepherded the library in its early years. Severson laughs as he recalls how Kirchfeld, now retired, began collecting books for the library.

"They couldn't afford a library budget, so Friedhelm was resourceful," Severson said. "He used to dumpster dive outside the library of the University of California, San Francisco (UCSF) campus. He collected a number of homeopathy journals that they discarded." Severson recently wrote to UCSF to tell its librarians how the ever-growing NCNM collection began with their discards—and now UCSF is sharing information with NCNM about digitalization.

"It's a small world," Severson said. "Friedhelm would be proud."

NCNM thanks Board of Regents member Priscilla Morehouse, who funded a reading area for the Mike and Simone Chilton to benefit all NCNM students. ▪

APPENDIX FOUR
WVSSA POSITION ON CANOLA IN THE WILLAMETTE VALLEY

Annex 4

February 13, 2013

WILLAMETTE VALLEY SPECIALTY SEED ASSOCIATION

WVSSA Opposition to Canola in the Willamette Valley

1. Membership has voted for no change to the last Oregon Department of Agriculture (ODA) administrative rule limiting canola (see sections 603-052-0850 through 603-052-0880). That rule protected the seed industry and was founded in well-rounded and inclusive discussions over the years. The ODA has adopted a destructive rule change without offering new evidence that would support the reversal of past decisions that were based on a scientific determination. By its own admission, the ODA is not allowed to make a decision to re-write its rules on a market basis.

2. Oregon law initiated a process through the ODA in 1989 that began a period of regulating canola that by 2005 in essence disallowed commercial production of canola in the Willamette Valley. It is not a coincidence that in the last 20-some years the Willamette Valley has taken full advantage of the unwise, unmanaged expansion of canola in Europe. Seed companies built strong businesses that grew remarkably while capturing the displaced production in Europe. It is thievery for the state of Oregon to reverse a policy of protecting specialty seed.

3. ODA has known all along that it is without evidence to support the reversal of past decisions based on science. ODA sent a letter on November 5, 2012, to the State Emergency Board in which the department requested $446,040 to research five issues. In summary, the research would investigate the potential for cross-pollination to damage related seed crops, assess canola volunteer persistence, study off-field movement of canola seed in waterways, evaluate baseline "feral" Brassicas in the area, and conduct a science-based policy, with market and risk analysis for coexistence of specialty seeds and canola. Clearly, the ODA has revised an existing, successful rule for which many substantial questions remain unanswered.

4. The economic impact of canola to the Willamette Valley is negative.

     a. The WVSSA is being asked to support or accommodate a low value commodity crop that will displace a portion of high value specialty seed crops, as well as putting at risk clover and grass seed shipped out of the state or overseas. Growing canola in the Willamette Valley will not add jobs to the Oregon economy. To the contrary, displacing jobs in high value seed crops will result in a net loss of jobs and income. Meanwhile, canola can be sourced from Oregon in production areas that are without specialty seed production.

     b. The farmgate value in recent years for seed interests negatively affected by a proliferation of canola is estimated as follows:

         i. Specialty seed: $50,000,000 in 2012.

         ii. Clover seed: $30,000,000 in 2012.

         iii. Grass seed over the last few years has declined, but still exceeded $228,000,000 in 2010. Over the last 10 years grass seed peaked at about $480,000,000 and is generally considered to have a value of about $300,000,000.

5. In order for the early proposals to the rule to function, the WVSSA would have had to modify seed production rules and guidelines and revise association Bylaws. This is excessive government intervention in a very successful business model that is world renowned. In essence, the WVSSA would have been forced to accommodate a non-seed crop in what is entirely a seed organization. Those proposals were unacceptable.

6. Canola proponents cite that this crop is needed in grass seed rotations, when in fact grass seed growers have testified that several other alternative crops exist. These alternatives typically require higher management or other input levels compared to canola, though this is hardly a reason to disqualify them.

7. The WVSSA contends that canola is damaging as a crop, weedy volunteer, and host to insect and disease pests negatively impacting specialty, clover, and grass seed producers and related interests, such as fresh market vegetables. Evidence is available. For example, Crucifer (Brassica species and radish) seed production areas in Europe have been ruined and lost indefinitely, particularly in Denmark, the UK, and France. This is long term damage that is not easily reversed or negated. Depending on the specific crop, canola variously affects seed or food quality. Examples include, but are not limited to, unwanted cross pollination, undesirable weed seed content, insect pests such as pollen beetles and cabbage root maggots, and plant diseases such as those caused by the *Sclerotinia* fungus.

8. The adopted rule seeks to balance the interest in canola not on a state-wide basis, but specifically in the Willamette Valley. This is a short-sighted form of coexistence, when a better form of coexistence is readily available by utilizing vast small grain growing areas to achieve balance. Certainly, Oregon grown canola can be produced in the state and supply oilseed processors without jeopardizing an existing industry.

MIKE CHILTON

After receiving a graduate degree in botanical sciences from Iowa State University, Mike spent the next fifteen years in Southeast Asia. He worked initially as an agricultural volunteer and team leader in Vietnam and then as a remote area security development specialist with the Thai Border Patrol. He returned to Vietnam to work in private and public sector development activities before returning to the USA in 1975 to an uncertain future and the development of a new career.

He served with a small agricultural company in Salem, Oregon, for eight years catching up on the world of agriculture before establishing his own company, Agricultural Alternatives, Ltd. in 1985. Specialty plant and seed production expanded into a broad range of plant products, all marketed internationally, before he sold the company in 2001.

Mike has visited and traveled in over eighty countries and has lived, worked or conducted business in nineteen countries. He lives with his wife, Simone, on an acreage near Salem, Oregon, and continues today to serve useful purposes following a number of ancillary interests.

BOB GRIFFIN

Bob began his career in international development with USAID in Laos in 1966. There his first job was as regional manager for southern Laos of a new joint venture of USAID and the Royal Lao Government called the Agricultural Development Organization. ADO was set up to provide agricultural inputs including tools, machinery, fertilizer and pesticides as well as credit and marketing services to Lao farmers. This work led to an attempt to develop a commercial rice plantation in Laos, a venture that was ultimately unsuccessful. Subsequently, he worked for a UNDP regional project that served UN country projects in regard to management, training and communications issues.

After three years, he became an independent consultant working of various types of international development projects but specializing in agricultural and rural development. He continued as an independent consultant contracted by various development agencies for the next thirty-five years, retiring in 2013.

Bob studied political science at Oberlin College and completed a master's degree in education at the University of Southern California and an MBA in international business at the City University of New York. He lives with his wife, Normalah, in Honolulu and enjoys spending time with his three adult daughters and three grandchildren.

CPSIA information can be obtained
at www.ICGtesting.com
Printed in the USA
FSOW01n1350141117
40958FS